John M. Swan

A Manual of Human Anatomy

Arranged for Second-Year Students

John M. Swan

A Manual of Human Anatomy
Arranged for Second-Year Students

ISBN/EAN: 9783337371470

Printed in Europe, USA, Canada, Australia, Japan

Cover: Foto ©berggeist007 / pixelio.de

More available books at **www.hansebooks.com**

A MANUAL

OF

HUMAN ANATOMY

ARRANGED FOR

SECOND YEAR STUDENTS

BY

JOHN M. SWAN, M. D.,

Assistant Demonstrator of Anatomy, University of Pennsylvania.

PHILADELPHIA

F. W. S. LANGMAID, M. D.

202 South 36th Street.

The references at the end of each section are to the following works:

A. T. O.—An American Text-Book of Obstetrics for Practitioners and Students. Edited By RICHARD C. NORRIS, M. D., Philadelphia, 1895.

GRAY.—Anatomy, Descriptive and Surgical. Edited By HENRY GRAY, F. R. S., New American Edition, Philadelphia and New York, 1897.

MORRIS.—Human Anatomy. Edited By HENRY MORRIS, M. A., M. B., Philadelphia, 1893.

PIERSOL.—Text-Book of Normal Histology. By GEORGE A. PIERSOL, M. D., Philadelphia, 1895.

QUAIN.—Quain's Elements of Anatomy. Edited By E. A. SCHÄFER, F. R. S., and GEORGE DANCER THANE. Tenth Edition, Vol. I., Part 1., London, 1896.

CHAPTER I.

EMBRYOLOGY.

The **ovum** is a large, round cell, $\frac{1}{125}$ inch (0.2 mm.) in diameter. It is derived from the germinal epithelium which covers the ovary.

The ovum consists of (1) the zona pellucida; (2) the vitelline membrane; (3) the vitellus, or yolk; (4) the germinal vesicle; and (5) the germinal spot.

(1). The **zona pellucida** is a protecting membrane which is derived from the cells of the discus proligerus.

(2). The **vitelline membrane** corresponds to the cell wall of an ordinary cell.

(3). The **vitellus** or yolk corresponds to the cell contents of an ordinary cell.

(4). The **germinal vesicle** corresponds to the nucleus of an ordinary cell.

(5). The **germinal spot** corresponds to the nucleolus of an ordinary cell. (Piersol, p. 227.)

The vitellus of an ovum is composed of the **ovaplasm,** or animal yolk, which is arranged in the form of a network, in the meshes of which we find the **deutoplasm,** or food yolk.

Ova are divided into three classes depending upon the relation of the deutoplasm to the ovaplasm:

1. **Alecithal;** ova in which the ovaplasm and deutoplasm are equally distributed throughout the vitellus; e. g., ova of man, mammals, amphibians, and the amphioxus.

2. **Telolecithal;** ova in which the deutoplasm is large in amount and is heaped up at one pole of the vitellus, while the ovaplasm is small in amount and lies above the deutoplasm; e. g., ova of birds, reptiles, and fishes.

3. **Centrolecithal;** ova in which the deutoplasm lies in the centre of the ovum, whilst the ovaplasm surrounds it as a peripheral covering; *e. g.*, ova of insects. (Quain, p. 8.)

In the ovary the ovum lies in a vesicle which is known as the **Graafian follicle.** The youngest Graafian follicles consist of an ovum, surrounded by a single layer of low columnar cells. A fully developed Graafian follicle consists of (1) the *theca folliculi*, a limiting membrane of condensed ovarian stroma; (2) the *membrana granulosa*, stratified, small, polyhedral cells, which are derived from the low columnar elements of the young follicle; (3) the *discus proligerus*, a thickening of the membrana granulosa; (4) the *ovum;* and (5) the *liquor folliculi*, an albuminous fluid filling the otherwise unoccupied portion of the follicle. (Piersol, pp. 225, 226.)

MATURATION. While the ovum lies in the Graafian follicle it undergoes a process which is termed maturation. This process is designed to prepare the ovum for the reception of the male element. Every ovum undergoes maturation, whether it is to become impregnated or whether it is to remain unimpregnated.

Process: 1. The germinal vesicle (nucleus) increases in size and moves toward the periphery of the ovum. 2. The nuclear membrane breaks down and a nuclear spindle appears. 3. The nuclear spindle projects against the vitelline membrane (cell wall), evaginates a small portion of the ovum, and a few chromasomes (filaments of chromatin) from the germinal vesicle pass into the portion thus evaginated. 4. The evaginated portion of the ovum becomes constricted off to form the **first polar body.** 5. The nuclear spindle again projects against the vitelline membrane, evaginates a second small portion of the ovum, into which a few more chromosomes pass. 6. The second evaginated portion becomes constricted off to form the **second polar body.**

Therefore, as a result of maturation, we have the formation of two polar bodies and the extrusion of a portion of the chromatin of the germinal vesicle. After the process of maturation is complete we find that the nucleus of the ovum returns to its former position; but it is now known as the **female pronucleus** or germ nucleus. (Quain, p. 9; A. T. O., p. 74.)

IMPREGNATION. Impregnation is accomplished by the union · of the ovum and one spermatic filament (spermatozoon). The **spermatic filament** is developed from the nucleus of one of the daughter cells which are found lining the seminiferous tubules in the testicle, and is, therefore, composed entirely of chromatin. The fully developed spermatic filament consists of a head, middle piece, and tail or cilium. (Piersol, pp. 209–211.)

The semen, containing numberless spermatic filaments, is deposited in the vaginal cul-de-sac, and, by the propelling force of the cilia, the spermatic filaments travel through the cervical canal, the cavity of the uterus, and the Fallopian tube, to meet the ovum. One spermatic filament, only, is required to impregnate an ovum. The spermatic filament which is to act as the impregnating body is received by the **receptive eminence** of the ovum, a protrusion of the protoplasm of the cell. The tail and middle piece of the spermatic filament disappear, whilst the head enters the substance of the ovum, after passing through the zona pellucida. From the head of the spermatic filament the **male pronucleus,** or sperm nucleus, is formed. The sperm nucleus and the germ nucleus then approach each other and lie in close apposition. (Quain, pp. 11–14; A. T. O., p. 75.)

It will be seen that up to this time the ovum has possessed three different nuclei; first, the *germinal vesicle;* second, the *female pronucleus,* or germ nucleus; and third, the *male pronucleus,* or sperm nucleus.

SEGMENTATION. As soon as the sperm and germ nuclei lie in close contact the ovum is ready to divide. The division of the ovum is known as segmentation, and takes place by the process of karyokinesis.[1] The first division takes place in a plane at right angles to the polarity of the original ovum: the resulting cells each possessing one-half the chromatin of the germ nucleus and one-half the chromatin of the sperm nucleus. This fact will explain the inheritance of characteristics of both parents. The two cells resulting from the first division of the ovum divide into four, the four divide into eight, the eight into sixteen, etc., until we have a mass of innumerable cells. The zona

[1] For Karyokinesis see Piersol, pp. 15-20.

pellucida does *not* divide; but surrounds the segmenting cells as a protecting membrane. (Quain, p. 16; A. T. O., p. 76.)

According to the character of segmentation we have ova classed as; (1) **holoblastic,** in which the entire cell divides; and (2) **meroblastic,** in which only a portion of the cell divides. Holoblastic ova may result in (*a*) **equal** segmentation, when the cells produced by the first division are of the same size; or in (*b*) **unequal** segmentation, when one of the cells resulting from the first division is much larger than the other. Meroblastic ova may result in (*c*) **discoidal** segmentation, when the cells resulting from the division of the ovum lie over the undivided portion as a disc; or in (*d*) **superficial** segmentation, when the cells resulting from the division of the ovum lie around the undivided portion in a superficial layer.

Varieties of Segmentation.

I. Holoblastic or Total. { *a.* Equal; *e. g.*, man, mammals, amphioxus. / *b.* Unequal; *e. g.*, amphibians, bony fishes.
II. Meroblastic or Partial. { *c.* Discoidal; *e. g.*, birds and reptiles. / *d.* Superficial; *e. g.*, insects.

As the result of the repeated division of the ovum we have a mass of cells surrounded, in the mammalian egg, by the zona pellucida. This structure is known as the **blastodermic vesicle.**

The **blastula,** or first stage of the blastodermic vesicle is composed of a layer of cells surrounding a central cavity, which contains a small amount of albuminous fluid, and which is known as the **segmentation cavity.** This stage was formerly known as the morula or mulberry mass. (Quain, p. 17.)

After the blastula is well formed, the cells at the lower pole of the blastodermic vesicle become invaginated and applied to the under surface of the cells at the upper pole. This stage is known as the **gastrula,** and results in the formation of a two-layered blastodermic vesicle; the outer layer of cells being called the **ectoderm,** the inner layer being called the **entoderm.** As a result of the invagination of the lower cells, the segmentation cavity becomes obliterated and we have the appearance of the **archenteron,** a cavity situated within the entodermic cells. The passageway from the archenteron out into the surrounding tissues is known as the **blastopore** and the cells bounding the blastopore on either side are known as the **lips** of the blastopore. These two stages have been carefully studied in the ovum of the amphioxus. (Quain, p. 21; A. T. O., pp. 77-79.)

The blastodermic vesicle of the mammalian ovum is surrounded by a single layer of cells derived from the zona pellucida, known as the **layer of Rauber.** Inside the layer of Rauber is a mass of cells which afterwards multiply and arrange themselves in two layers to form the ectoderm and entoderm. (A. T. O., p. 78.)

Growing between the ectoderm and entoderm we find a third layer of cells which are derived from the multiplication of the entoderm, principally, and of the ectoderm to a slight extent; this is the **mesoderm.** In the lower types the mesoderm begins to form at the lips of the blastopore. (A. T. O., p. 79.)

PRIMARY EMBRYONIC FORMATIONS. (1) The primitive streak; (2) the medullary folds and medullary groove; (3) the notochord; and (4) the mesodermic somites.

1. THE PRIMITIVE STREAK. In the embryonic area of the developing chick ovum we may distinguish an inner, clear portion, area pellucida, and an outer, obscure portion, area opaca. The latter portion lies on the food yolk. At the junction of the area pellucida and the area opaca, posteriorly, we may see a semilunar fold of tissue which is called the embryonic sickle or the **primitive crescent.** From the embryonic sickle the primitive streak grows forward into the area pellucida. This formation has nothing to do with the embryo and is formed by the fusion of the lips of the blastopore. The theory which thus accounts for the formation of the primitive streak is the **concrescence theory** of Minot. (A. T. O., p. 78.)

2. THE MEDULLARY FOLDS. In front of the primitive streak the ectoderm undergoes rapid proliferation and folding takes place. The folds are the medullary folds, and between them is the medullary groove. This is the first indication of the embryo. As the growth progresses the medullary folds approach each other

and unite over the dorsum of the embryo, converting the medullary groove into a canal, the **neural canal.** The closure of the medullary groove takes place first in the cervical region, and progresses thence toward the head and toward the tail. The brain and spinal cord are developed from the walls of the neural canal, while the canal itself becomes the central canal of the spinal cord and the ventricles of the brain. (Quain, pp. 30-32; A. T. O., p. 80.)

3. THE NOTOCHORD is derived from the entoderm, and lies just beneath the neural canal. It is the first indication of the axis of the embryo. The notochord persists in the amphioxus; in man it is represented by the pulpy substance in the intervertebral disc. (Quain, p. 32; A. T. O., p. 81.)

4. THE MESODERMIC SOMITES. The upgrowth of the notochord and the downgrowth of the neural canal divide the mesoderm into two parts. By a process of vertical cleavage, the mesoderm on either side of the axis of the embryo is divided into two parts; the **para-axial mesoderm,** lying in close relation with the axis of the embryo; and the **peripheral mesoderm,** lying remote from the axis. In the paraaxial mesoderm, by a process of horizontal cleavage, twenty-four quadrilateral plates of mesodermic tissue are formed which are known as the **mesodermic somites.** The upper and outer portions of each one of the somites form the **muscle plate,** from the cells of which all the voluntary muscle in the body is developed. After the formation of the muscle plate the remainder of the somites fuse, and from the resulting tissue the true vertebræ are formed in such a way that one vertebra is developed from the adjacent halves of two somites. (A. T. O., pp. 81-83.)

The **celom** or primitive pleuro-peritoneal cavity is developed in the peripheral mesoderm by a process of liquefaction. The result of the appearance of the body cavity is the division of the peripheral mesoderm into two layers. The outer or parietal layer joins with the ectoderm to form the **somatopleure,** while the inner or visceral layer joins with the entoderm to form the **splanchnopleure.** (A. T. O., pp. 79 and 82.)

The **primitive gut** is formed by the anterior folding of the splanchnopleure and is, therefore, lined by entoderm and covered on its outer surface by mesoderm. At first the gut is widely open and communicates with a sac which, in the mammal, is known as the **umbilical vesicle.** As the gut becomes closed the communication between it and the umbilical vesicle becomes more and more narrow, until there is merely a duct passing through the body wall between the two structures. This duct is contained in the **abdominal** or **allantoic stalk,** which also carries the vitelline vessels, that have developed in the mesoderm of the umbilical vesicle, into the body of the embryo. The vitelline sac of the chick corresponds to the umbilical vesicle of the mammal. (A. T. O., p. 112.)

The **body wall** is formed by the anterior folding of the somatopleure.

By the term **mesothelium** we understand the mesodermic cells which limit the pleuro-peritoneal cavity. These cells are the direct ancestors of the endothelium.

FETAL MEMBRANES. The fetal membranes are the *amnion,* the *allantois,* and the *chorion.* Animals may be classed as **amniota,** when they are provided with an amnion during development, and as **anamnia,** when they are developed without the formation of an amnion.

The **amnion** is formed by a folding of the somatopleure over the dorsum of the embryo. The amnion folds, and the folds which afterwards meet to form the body wall, are continuous, both growing at the same time. The amnion begins to fold upward simultaneously from the head, from the tail, and from the sides. As these folds grow they turn upon themselves and lie beneath the zona pellucida. The fold of somatopleure nearer the embryo is known as the **true amnion** and is lined by ectoderm. The reflected fold, which lies beneath the zona pellucida, is known as the **false amnion** and has ectoderm on its outer surface. By the **amniotic suture** we mean the line of union of the head, tail, and lateral amnion folds. The sac formed between the embryo and the amnion contains the **amniotic fluid.** The amniotic fluid serves to protect the fetus from injury, to maintain an equable temperature, and to provide the requisite amount of fluid for the tissues. (Quain, p. 42; A. T. O., pp. 83-84.)

The **allantois** grows from the hind gut, and is, therefore, lined by entodermic cells. As the allantois grows the allantoic vessels are carried out in its mesodermic tissue. In the developing chick the allantois is a free sac, which lies beneath the air chamber of the egg, and, in addition to serving as a receptacle for effete matter, is a respiratory apparatus. In the human embryo we never have a free allantois; the membrane grows with great rapidity and joins with other structures to form the true chorion. It is connected to the body of the embryo by the allantoic or **abdominal stalk.** (Quain, pp. 43-46; A. T. O., p. 84.)

The zona pellucida, which surrounds the developing ovum as a protecting membrane, is known as the **prochorion.** The prochorion and false amnion fuse to form the **primitive chorion.** The primitive chorion and the allantois fuse to form the **true chorion.** The true chorion is covered on its free surface by numerous club-shaped villi and these villi are in turn covered by ectodermic cells derived from the false amnion. These villi atrophy over a portion of the chorion, which is termed the **chorion leve;** whilst they persist over the remainder of the membrane, which is known as the **chorion frondosum.** Branches of the allantoic blood vessels grow into the villi of the chorion frondosum. The presence of chorionic villi in a vaginal discharge is a positive indication of pregnancy. (Quain, pp. 42-46; A. T. O., p. 84.)

CHANGES IN THE UTERINE MUCOUS MEMBRANE. Every twenty-eight days during the sexual life of the female, the uterine mucous membrane undergoes a cycle of changes, which results in the formation of the decidua, when an ovum becomes impregnated; or which results in the phenomena of menstruation, when impregnation does not occur. The changes in the appearance of the mucous membrane of the uterus of the non-pregnant female may be described in four periods: first, a period of **preparation,** which lasts seven days; second, a period of **degeneration,** which lasts five days; third, a period of **repair,** which lasts four days; and fourth, a period of **rest,** which lasts twelve days. In case an ovum becomes fecundated, the second period, or the period during which we see the active phenomena of menstruation, fails to follow the first period in the cycle and we see the ovum received in a fold of the hypertrophied mucous membrane and there lodged. The hypertrophy of the uterine mucous membrane then progresses and we have the **decidua** formed. The **decidua reflexa** is that portion of the uterine mucous membrane which is thrown around the ovum, holding it in place. The **decidua serotina** is that portion of the uterine mucous membrane which is situated between the ovum and that portion of the uterine wall to which the ovum becomes attached. The **decidua vera** is that portion of the uterine mucous membrane which lines the true uterine cavity. (Quain, pp. 46-53; A. T. O., p. 86.)

The **placenta** is the structure by which the fetus is attached to the uterus and through which the fetus receives its nourishment and its oxygen. The human placenta is formed partly from fetal and partly from maternal tissue. The fetus contributes amnion and chorion frondosum and the mother contributes the decidua serotina to the formation of this structure. The villi of the chorion frondosum are received into the dilated capillary blood vessels of the decidua serotina. There is no actual mingling of the maternal and fetal blood, a thin membrane separating the two currents.

$$\text{Placenta} \begin{cases} \text{Fetal} \begin{cases} \text{Amnion.} \\ \text{Chorion frondosum.} \end{cases} \\ \text{Maternal} \begin{cases} \text{Decidua serotina.} \end{cases} \end{cases}$$

At the termination of gestation the membranes which have enclosed the fetus are thrown off. From within outward we find (1) the amnion, (2) the chorion leve, (3) the decidua reflexa, (4) the decidua vera. (Quain, pp. 53-55; A. T. O., pp. 86-93.)

The fetus is connected with the placenta by the **umbilical cord,** which is composed of (1) a cleft corresponding to the body cavity, (2) the umbilical stalk, (3) the vitelline vessels (two arteries and two veins), (4) the allantoic stalk, (5) the umbilical vessels (two arteries and one vein), (6) the jelly of Wharton, and (7) the amnion. (A. T. O., pp. 93-94.)

THE VISCERAL ARCHES. On either side of the head of the developing embryo five arches of tissue may be seen which are separated from each other by four furrows. The arches are the so-called **visceral arches** and the furrows are the **visceral furrows.** Each of the visceral arches contains a rod of cartilage and one of the aortic arches. The first visceral arch bifurcates into a superior and an inferior process. The superior process becomes the superior maxilliary bone; the inferior process becomes the inferior maxillary bone. From the rod of cartilage contained in the first visceral arch, which is called Meckel's cartilage, the malleus, the incus and the stylo-maxillary ligament are developed.

The second visceral arch, by its contained rod of cartilage, form the stapes, the styloid process of the temporal bone, the stylo-hyoid ligament, and the lesser cornu of the hyoid bone.

The third visceral arch, by its contained rod of cartilage, forms the greater cornu of the hyoid bone.

The fourth and fifth visceral arches fuse to form the tissues of the neck.

The first visceral furrow becomes the external auditory meatus.

The second, third and fourth visceral furrows disappear.

The **external ear** is developed in the tissues of the first and second visceral arches, around the first visceral furrow. (A. T. O., p. 96.)

THE PHARYNGEAL POUCHES. If we examine the pharynx of a developing embryo, we will find that there are four pharyngeal pouches which correspond in position with the four external visceral furrows.

The first pharyngeal pouch becomes the tympanum and the Eustachian tube.

The second pharyngeal pouch disappears.

The third pharyngeal pouch forms the thymus gland.

The fourth pharyngeal pouch forms the lateral portion of the thyroid gland. (A. T. O., p.113.)

THE FACE. The **naso-frontal** process grows down from the head and forms the bridge of the nose. The **lateral process** is a knob-like mass of tissue situated by the side of the naso-frontal process and continuous with it above; but separated from it below by a notch. The notch becomes the nostril; the lateral process forms the ala of the nose.

The groove between the lateral process and the superior process of the first visceral arch leads to the developing eye and becomes the **nasal duct.** (A. T. O., p. 97.)

THE EXTREMITIES. The arms appear at about the twenty-third day as bud-like processes from the upper thoracic region. The legs appear a few days later from the sacral region. The first segment to appear is the hand or foot, as the case may be. About a week later the forearm and leg appear, and in another week the arm or thigh may be seen. At this time the digits are differentiated.

The flexures begin at about the twenty-first day. There are four flexures; the cephalic, the cervical, the dorsal, and the sacral. The flexures are most pronounced at about the twenty-third day, and after this period the fetus gradually unwinds. (A. T. O., p. 97.)

The development of the fetus may be divided into three stages: first, the blastodermic stage, from the first to the twelfth day; second, the embryonal stage, from the thirteenth to the twenty-eighth day; and third, the fetal stage, from the beginning of the fifth week to the completion of gestation. (A. T. O., p. 94.)

Haase's rule for the determination of the age of the fetus. During the first five months the age of a given fetus will be the square root of its length. During the second five months the age of a given fetus will be its length divided by five, $e.$ $g.$, a fetus 16 cm. long is 4 months old ($\sqrt{16}=4$); a fetus 30 cm. long is 6 months old ($30 \div 5 = 6$). (A. T. O., p. 103.)

NOTE:—The embryology of the nervous system, the cardio-vascular system, etc., will be found in the chapters treating of the anatomy of the respective parts.

CHAPTER II.

THE CENTRAL NERVOUS SYSTEM.

The central nervous system is composed of (1) the cerebrum, (2) the cerebellum, (3) the crura cerebri, (4) the pons Varolii, (5) the medulla oblongata, and (6) the spinal cord.

THE SPINAL CORD.

The **spinal cord** is a cylindrical mass of nervous tissue, eighteen inches in length, which is contained in the vertebral canal. The spinal cord begins at the upper margin of the foramen magnum and terminates, at the lower border of the first lumbar vertebra, in a cone-shaped end which is called the **conus medullaris.** (Morris, p. 757; Gray, p. 695.)

As the spinal cord lies in the vertebral canal, it is surrounded by the spinal meninges or membranes. The membranes enveloping the spinal cord are three in number; the *dura mater,* the *arachnoid,* and the *pia mater.* The **dura mater** or outer membrane is a strong, fibrous, protecting membrane. The **arachnoid** lies immediately beneath the dura mater, being separated from it by a capillary lymph space, the **sub-dural lymph space.** Between the arachnoid and the underlying pia mater there is a considerable space, the **subarachnoid space,** which is occupied by the **cerebro-spinal fluid.** Crossing the subarachnoid space we see the anterior and posterior roots of the spinal nerves, the ligamentum denticulatum, and the septum posticum. The **septum posticum** is a process of the arachnoid which carries bloodvessels to the pia mater. The **pia mater** is a vascular membrane which invests the spinal cord closely, and which is prolonged into the white matter of the cord as connective tissue septa. The **ligamentum denticulatum** is a process of the pia mater which passes across the subarachnoid space between the anterior and posterior roots of the spinal nerves, connecting the pia mater with the dura mater. The **filum terminale** is a process of the pia mater, which extends from the conus medullaris to be attached to the base of the coccyx. In its upper portion the filum terminale contains a small amount of nervous matter. (Morris, pp. 754–757; Gray, p. 693.)

The spinal cord possesses two fissures; a true, **anterior fissure,** and a **posterior fissure,** which is merely a septum of pia mater prolonged into the cord.

The spinal cord is larger in the cervical and in the lower thoracic regions than it is in the upper thoracic region. These enlargements are known, respectively, as the cervical and lumbar enlargements. The **cervical enlargement** extends from the third cervical to the second thoracic vertebra. The **lumbar enlargement** extends from the ninth thoracic to the twelfth thoracic vertebra. The enlargements may be accounted for by the exit of the roots of the cervical, lumbar, and sacral nerves, which enter into the formation of the cervical, brachial, lumbar, and sacral plexuses.

On the lateral aspect of the spinal cord the roots of the spinal nerves are to be seen as they leave and enter the nervous tissue. The anterior roots are motor and the posterior roots are sensory in function. The posterior root is furnished with a ganglion. Below the conus medullaris the roots of the lumbar and sacral nerves, passing downward to their foramina of exit from the vertebral canal, form a bundle which is known as the **cauda equina.** The cauda equina is surrounded by the arachnoid and by the dura mater. (Morris, pp. 757–759; Gray, p. 696.)

The white matter of the spinal cord is composed of: (1) medullated nerve fibres, (2) neuroglia, (3) bloodvessels, (4) lymphatics, and (5) ingrowths of the pia mater.

The white matter is divisible into the anterior, the lateral, and the posterior columns. The anterior column is situated between the anterior fissure and the anterior horn of gray matter; the lateral column is situated between the anterior and the posterior horns of the gray matter; and the posterior column is situated between the posterior fissure and the posterior horn of gray matter.

The white matter of the spinal cord is composed of various tracts of fibres, the functions of which are more or less completely understood:

1. **The anterior** (direct) **pyramidal tract** is situated in the anterior column of white matter, in close relation with the anterior fissure.

2. **The lateral** (crossed) **pyramidal tract** is situated in the lateral column of white matter, alongside the posterior horn of gray matter. This tract does not occupy the entire thickness of the lateral column.

3. **The direct cerebellar** tract is situated in the lateral column of white matter, between the lateral pyramidal tract and the periphery of the cord.

4. **The column of Goll** is situated in the posterior column of white matter, alongside the posterior fissure.

5. **The column of Burdach** is situated in the posterior column of white matter, alongside the posterior horn of gray matter.

6. **The ascending antero-lateral tract** (column of Gowers) is situated in the lateral column of white matter, along the periphery of the cord and anterior to the direct cerebellar tract.

7. **The descending antero-lateral tract** (column of Löwenthal) is situated in the lateral column of white matter, along the periphery of the cord, anterior to the ascending antero-lateral tract. This tract extends across the anterior nerve roots as far as the anterior pyramidal tract.

8. **The mixed lateral tract** is situated in the lateral column of white matter, alongside the anterior horn of gray matter.

9. **The antero-lateral ground bundle** is situated in the lateral tract of white matter. It extends from the anterior extremity of the lateral pyramidal tract to the white commissure. In its course it lies between the ascending antero-lateral tract and the mixed lateral tract; between the descending antero-lateral tract and the anterior horn of gray matter; and between the anterior horn of gray matter and the anterior pyramidal tract.

10. **The column of Lissauer** lies between the posterior horn of gray matter and the periphery of the cord. (Morris, pp. 762-765; Gray, p. 699.)

The gray matter of the spinal cord is composed of two lateral masses of nervous substance which are connected with each other by the **gray commissure.** The gray commissure lies immediately beneath the posterior fissure and is separated from the anterior fissure by the **white commissure,** which connects the white matter of the two sides of the cord. The gray commissure contains the **central canal** of the spinal cord, which is the remains of the neural canal of the embryo. In the adult, the central canal of the spinal cord is lined by ciliated columnar epithelium. Each half of the gray matter may be divided into a larger, **anterior horn** and a smaller, **posterior horn.**

Histologically, the gray matter is composed of (1) nerve cells, (2) non-medullated nerve fibres, (3) neuroglia, (4) blood-vessels, and (5) lymphatics.

The nerve cells in the anterior horn of gray matter are principally large, multipolar, ganglion cells. These cells are arranged as a **mesial group,** near the inner aspect of the horn; an **anterior group,** and a **lateral group. The column of Clarke** is a column of nerve cells situated at the junction of the posterior horn with the gray commissure. These cells are seen only in the thoracic portion of the cord. (Morris, p. 761; Gray, p. 701.)

The neuroglia is condensed over the posterior horn of gray matter, to form the **substantia gelatinosa Rolandi;** and around the central canal, to form the **substantia gelatinosa centralis.**

The **anterior roots of the spinal nerves** are composed of fibres which are the neurits of cells in the anterior horn of gray matter. Some of the cells in the anterior horn of gray matter are innervated by the terminal arborizations of neurits which have come from cells in the cerebral cortex, around the fissure of Rolando, through the anterior and lateral pyramidal tracts of the spinal cord. Some of these fibres decussate in the medulla, at the decussation of the pyramids, and others in the cord, by passing through the white commissure of the cord. Others of the cells in the anterior horn of gray matter are innervated by the terminal arborizations of neurits which have come from the cells in the posterior horn of gray matter.

The cells which control the sensory nerve fibres are situated in the spinal ganglia. One of these cells sends out a process which divides at once into two branches: one of which passes to the periphery of the body as a sensory nerve fibre, and the other of which passes through the posterior nerve root and enters the cord. This latter fibre divides into a branch which passes up the cord and a branch which passes down the cord, in the white matter. These branches in turn send off collateral branches, which enter the gray matter at different levels.

The posterior roots of the spinal nerves are composed (1) of fibres which enter the column of Burdach, (2) of fibres which enter the column of Lissauer and thence pass to form terminal aborizations around the cells in the posterior horn of gray matter, and (3) of fibres which enter the substantia gelatinosa Rolandi. Of the fibres which enter the column of Burdach, a first group passes into the column of Goll of the same side and then runs up and down the cord; a second group passes into the column of Goll of the opposite side and then runs up and down the cord; a third group passes into the gray matter, forms aborizations around the cells in the column of Clarke, which cells, in turn, send neurits through the direct cerebellar tract to the cerebellum; a fourth group passes into the gray matter, forms arborizations around the cells in the posterior horn, which cells, in turn, send neuritis to form terminal aborizations around the cells in the anterior horn. (Morris, p. 762; Gray, p. 700.)

ANTERIOR ROOT.

I. Fibres from cells in anterior horn of gray matter of the same side ;
  a, mesial group,
  b, anterior group,
  c, lateral group.
II. Fibres from cells in anterior horn of gray matter of opposite side.

POSTERIOR ROOT.

I. Fibres enter the column of Burdach and thence pass ;
  a, to the column of Goll of same side ;
  b, to the column of Goll of opposite side ;
  c, to cells in the column of Clarke ;
  d, to cells in the posterior horn of gray matter.
II. Fibres enter the column of Lissauer.
III. Fibres enter the substantia gelatinosa Rolandi.

THE BRAIN.

All that portion of the central nervous system contained in the skull is known, collectively, as the brain. The brain is composed of (1) the medulla oblongata, (2) the pons Varolii, (3) the cerebellum, (4) the crura cerebri, and (5) the cerebrum.

Man has absolutely the heaviest brain of all animals, except the whales and the elephants. Relatively, man's brain is heavier than that of any other animal; being one forty-fifth the body weight or about fifty ounces.

The **cerebral meninges** are the membranes which cover the brain and are three in number; the dura mater, the arachnoid, and the pia mater.

The **dura mater** is a fibrous, protecting membrane and presents the following three points of difference from the dura mater of the spinal cord: first, the dura mater of the brain sends processes into certain fissures of the brain; the dura mater of the spinal cord does not send such processes into the fissures of the cord. Second, the dura mater of the brain forms the periosteum of the cranial bones; the dura mater of the spinal cord does not form the periosteum of the bones of the vertebral canal. Third, the dura mater of the brain contains venous sinuses; the dura mater of the spinal cord does not contain sinuses.

The **falx cerebri** is a process of the dura mater which projects into the longitudinal fissure of the brain. It is attached to the crista galli, to a ridge on the under surface of the frontal bone, to the under surface of the sagittal suture, to the superior arm of the occipital cross, and to the superior surface of the tentorium cerebelli. The unattached margin lies just above the corpus callosum.

The **tentorium cerebelli** is a process of the dura mater which projects into the fissure which separates the cerebrum, above, from the cerebellum, below. The tentorium cerebelli is attached to the lateral arms of the occipital cross, to the superior borders of the petrous portions of the temporal bones, and to the posterior and anterior clinoid processes. It makes an incomplete septum across the cavity of the cranium, through the opening in which the crura cerebri pass, from the cerebrum to the pons Varolii.

The **falx cerebelli** is a process of dura mater which projects into the fissure between the two lateral masses of the cerebellum. It is attached to the inferior arm of the occipital cross. (Morris, p. 694; Gray, p. 703.)

The dura mater of the brain sends processes along the cranial nerves as they pierce it to pass through the foramina of exit from the skull. These processes of dura mater become continuous with epineurium of the nerves.

The dura mater is supplied by the following arteries: the middle meningeal and the small meningeal, branches of the internal maxillary; the anterior meningeal, a branch of the internal carotid; and the meningeal branches of the occipital, ascending pharyngeal, vertebral, and anterior and posterior ethmoidal arteries. (Morris, p. 697; Gray, p. 703.)

There are fifteen venous sinuses situated between the layers of the dura mater. These sinuses drain the blood from the brain, and, partially, from the meninges. Five of the sinuses are arranged in pairs; (1) the **cavernous sinuses,** on either side of the body of the sphenoid bone; (2) the **superior petrosal sinuses,** along the superior borders of the petrous portions of the temporal bones; (3) the **inferior petrosal sinuses,** along the lower borders of the petrous portions of the temporal bones; (4) the **lateral sinuses,** beginning at the torcula Herophili, passing along the lateral arms of the occipital cross, and passing in a curved manner (sigmoid sinus) across the mastoid portion of the temporal bone; and (5) the **occipital sinuses,** beginning on either side of the foramen magnum and passing upward on either side of the inferior arm of the occipital cross. In some subjects there will be but one occipital sinus found. Five of the sinuses are single: (1) the **superior longitudinal sinus,** in the bony attachment of the falx cerebri; (2) the **inferior longitudinal sinus,** in the free border of the falx cerebri; (3) the **straight sinus,** in the attachment of the falx cerebri to the tentorium cerebelli; (4) the **circular sinus,** connecting the wo cavernous sinuses, and lying in the tissue covering in the sella turcica; and (5) the **transverse sinus,** lying across the basilar process of the occipital bone. The cavernous sinus of either side receives the blood from the ophthalmic vein, and bifurcates into the superior petrosal and the inferior petrosal sinuses. The superior petrosal sinus empties into the lateral sinus, which afterwards passes out of the skull through the jugular foramen. The inferior petrosal sinus leaves the skull by passing through the jugular foramen, and unites with the lateral sinus to form the internal jugular vein.

The straight sinus is formed by the union of the inferior longitudinal sinus and the veins of Galen.

The **veins of Galen** receive the blood from the choroid plexus of the lateral ventricle and join the inferior petrosal sinus iust as they leave the velum interpositum.

The **torcula Herophili** is situated in front of the internal occipital protuberance. It is the point of meeting of six sinuses; (1) the *superior longitudinal,* (2) the *straight,* (3 and 4) the two *lateral,* and (5 and 6) the two *occipital.* (Morris, p. 643; Gray, p. 657.)

The **arachnoid** is a thin membrane which lies between the dura mater and the pia mater. It does not dip down into the fissures of the brain, except into those which are occupied by the processes of dura mater previously mentioned. The space between the arachnoid and the pia mater is occupied by delicate trabeculæ of **subarachnoid tissue,** between the filaments of which we find the **cerebro-spinal fluid.** The **Pacchionian bodies** are hypertrophies of the subarachnoid tissue, which come in relation with the superior longitudinal sinus or with the bones of the calvarium. The **cisterna magna** is a dilatation of the subarachnoid space lying beneath the cerebellum and above the roof of the fourth ventricle. This space communicates, through the roof of the fourth ventricle, with the ventricular cavities of the brain, by the **foramen of Majendie.** The **cisterna pontis** is a dilatation of the subarachnoids pace, situated in front of the pons and behind the dorsum sellæ. This space is just in front of the circle of Willis. (Morris, p. 699; Gray, p. 704.)

The **pia mater** is the vascular membrane of the brain and

covers the brain substance closely, dipping down into all the fissures. It carries the bloodvessels which supply the brain. The brain is supplied by the two vertebral and the two internal carotid arteries and their branches. At the base of the brain the branches of these vessels form a free anastomosis, which is known as the circle of Willis. The **circle of Willis** is formed by the tip of the basilar artery; the two posterior cerebral arteries, branches of the basilar; the two posterior communicating arteries, branches of the internal carotid arteries; the tips of the two internal carotid arteries; the two anterior cerebral arteries, branches of the internal carotid arteries; and the anterior communicating artery, which connects the two anterior cerebral arteries. The posterior communicating artery connects the internal carotid and the posterior cerebral arteries. (Morris, pp. 526 and 701; Gray, pp. 573 and 705.)

THE MEDULLA OBLONGATA.

The **medulla oblongata** is that portion of the central nervous system situated just above the spinal cord and just below the pons Varolii; it ends at the upper margin of the foramen magnum. The medulla may be examined either from its anterior or from its posterior aspect. Anteriorly we see an **anterior median fissure**, which terminates just beneath the pons in the **foramen cecum.** On either side of the anterior fissure there is a tract of fibres known as the **pyramid,** which, in its lower portion, may be seen sending off trabeculæ, which cross the anterior fissure, forming the **decussation of the pyramids.** In the lower portion of the medulla the lateral tract is situated just external to the pyramid; but, in the upper portion of the organ, this tract becomes pushed backward by the appearance of a new body, the inferior olive. The inferior olive contains a nucleus of gray matter, known as the **dentate** or **olivary nucleus,** which is open toward the median line by its **hilum.** Just below the inferior olive we see an arching band of fibres which are the **anterior superficial arcuate fibres.** Therefore, on the anterior surface of the medulla we see three tracts on either side of the anterior fissure; (1) the pyramid, (2) the lateral tract, and (3) the inferior olive.

Posteriorly, we see that the medulla, in its upper half, is deflected from the median line, and that it passes over towards the cerebellum as two large bundles of nerve fibres known as the **restiform bodies** or the inferior peduncles of the cerebellum. In the lower half of the posterior surface of the medulla we may observe a **posterior median fissure** bounded on either side by the **funiculus gracilis.** Outside the funiculus gracilis we see the **funiculus cuneatus** and outside the latter tract we see the **funiculus of Rolando.** The funiculus of Rolando is limited externally by the lateral tract which was seen on the anterior surface of the organ. The funiculus gracilis and the funiculus cuneatus each present a decided enlargement in the upper portions of their courses. The thickening on the funiculus gracilis is called the **clava,** while that on the funiculus cuneatus is termed the **cuneate tubercle.** Both these thickenings are caused by under-lying collections of gray matter. The funiculus of Rolando contains the **nucleus of Rolando.** The restiform body is formed, macroscopically, by the continuations of the funiculus gracilis, the funiculus cuneatus, the funiculus of Rolando, and the lateral tract of the medulla. It passes to the cerebellum as the inferior peduncle of the cerebellum. Therefore, on the posterior surface of the medulla we see four tracts; (1) the funiculus gracilis, (2) the funiculus cuneatus, (3) the funiculus of Rolando, and (4) the restiform body. (Morris, p. 743; Gray, p. 708.)

If we make sections of a medulla which has been stained by Weigert's method, the microscopic appearance at the decussa-tion of the pyramids is quite different from the appearance at a plane passing through the inferior olive. At the decussation of the pyramids we find, passing through the section, a central **median raphé.** In front of this we see the fibres contained in the pyramid passing across the anterior fissure to take up their position in the lateral column of the spinal cord. On their way to the lateral column of the cord these fibres pass across the anterior horn of gray matter, isolate it, and lie for a short distance in the lateral tract of the medulla. The isolated anterior horn of gray matter is afterwards known as the **nucleus lateralis.** Posterior to the decussation of the pyramids, we find other fibres crossing the raphé. This is the beginning of the sensory decussation which takes place throughout the entire extent of the medulla and of the pons. Some of these fibres, after they have crossed the raphé, bend sharply at right angles and pass upward toward the brain. These are the **internal arcuate fibres.** In this situation the posterior columns are beginning to increase in size and the gray matter may be seen making its appearance among the nerve fibres.

At the level of the inferior olive, we see the pyramids, in front. External to the pyramid, on either side, is the inferior olive with its contained nucleus of gray matter, the **dentate nucleus.** The hilum of the dentate nucleus is open toward the median line and through the hilum countless nerve fibres pass. The **formatio reticularis** is seen in the centre of the section. This is composed; (1) of fibres which are crossing (internal arcuate fibres), cut longitudinally, (2) of fibres which have crossed at a lower point and which are passing toward the brain, cut transversely, and (3) of fibres which are passing toward the brain from the antero-lateral tract of the cord, cut transversely. The third group of fibres is spoken of as the **posterior longitudinal bundle.** In the posterior region of the medulla the nucleus gracilis and the nucleus cuneatus are seen. The fibres in the funiculus gracilis and the funiculus cuneatus form terminal arborizations around the cells in these nuclei and the neurits from these cells pass to various other parts of the nervous system.

The **internal arcuate fibres** are fibres which come from cells in the nucleus gracilis and nucleus cuneatus, cross the raphé in the substance of the medulla, and then pass upward toward the brain.

The **posterior superficial arcuate fibres** are fibres which come from cells in the nucleus gracilis and nucleus cuneatus, which go to the restiform body of the same side, and thence to the cerebellum.

The **anterior superficial arcuate fibres** are fibres which come from cells in the nucleus gracilis and nucleus cuneatus and which pass across the raphé, out of the anterior median fissure of the medulla, beneath the inferior olive, to the restiform body of the opposite side.

The **restiform body** is composed of fibres from the following sources.—

1. From the direct cerebellar tract of the same side.

2. From the nucleus gracilis and nucleus cuneatus of the same side (posterior superficial arcuate fibres).

3. From the nucleus gracilis and nucleus cuneatus of the opposite side (anterior superficial arcuate fibres).

4. From the olivary nucleus by fibres which pass;

        a, through the opposite olive;

        b, superficial to the opposite olive;

        c, deeper than the opposite olive.

5. From the superior olive.

6. From the lateral nucleus.

(Notice the difference between the fibres constituting the restiform body and the macroscopic formation of the same structure.)

The pyramids of the medulla are composed of fibres which occupy the anterior pyramidal and lateral pyramidal tracts of the spinal cord.

The lateral tract of the medulla is composed of fibres which occupy the antero-lateral tract of the spinal cord, and of the fibres which occupy the lateral pyramidal tract of the spinal cord on their way to the lateral column.

The funiculus gracilis of the medulla is composed of fibres which occupy the column of Goll in the spinal cord.

The funiculus cuneatus is composed of fibres which occupy the column of Burdach in the spinal cord. (Piersol, pp. 295-301; Morris, pp. 743-749; Gray, pp. 712-719.)

THE PONS VAROLII.

The **pons Varolii** is a mass of nervous tissue which is situated just above the medulla oblongata. The pons rests on the dorsum ephipii of the sphenoid bone. On its anterior surface the pons appears to be composed of transverse fibres. On section, however, we see that the organ is divisible into an anterior portion, or **crusta,** and a posterior portion, or **tegmentum.**

The **tegmentum** contains a continuation of the formatio reticularis, the fillet, and the superior olive. The **superior olive** is a collection of gray matter.

The **crusta** contains fibres which are passing transversely, between the two hemispheres of the cerebellum, and fibres which are passing longitudinally from the crura cerebri into the medulla. The transverse fibres form the **middle peduncle of the cerebellum.** A prominent group of these transverse fibres, situated just in front of the tegmentum, is known as the **trapezium.** The longitudinal fibres are found in the pyramids of the medulla. (Piersol, pp. 301-303; Morris, p. 742; Gray, p. 719.)

THE CEREBELLUM.

The **cerebellum** is a mass of nervous matter situated beneath the posterior lobe of the cerebrum, to the side of the pons, and above the medulla. It consists, macroscopically, of the **vermiform process** and two **hemispheres.**

Lobes of the cerebellum. The cerebellum is divided into a superior and an inferior surface by the **great horizontal fissure.** The vermiform process and the hemispheres are subdivided by secondary fissures into lobules. The lobules of the hemispheres bear different names from the corresponding lobules of the vermiform process. Certain of these lobules, however, are continuous from side to side, from one hemisphere through the vermiform process to the other hemisphere, to form lobes. On the superior surface, proceeding from before backward, we find: the **lingula,** continuous laterally with the **frenula,** to form the **lobus lingualis;** the **lobulus centralis,** continuous laterally with the **alæ,** to form the **lobus centralis;** the **culmen,** continuous laterally with the **anterior crescentic lobules,** to form the **lobus culminis;** the **clivus,** continuous laterally with the **posterior crescentic lobules,** to form the **lobus clivi;** the **folium cacuminis,** continuous laterally with the **posterior superior lobules,** to form the **lobus cacuminis.** On the inferior surface of the cerebellum, passing from behind forward, we find: the **tuber valvulæ,** continuous laterally with the **posterior inferior lobules** and with the **slender lobules,** to form the **lobus tuberis;** the **pyramid,** continuous laterally with the **biventral lobules,** to form the **lobus pyramidalis;** the **uvula,** continuous laterally with the **amygdalæ** to form the **lobus uvulæ;** and the **nodule,** continuous laterally with the **flocculi** to form the **lobus nodulæ.**

The **lingula** rests on the **superior medullary velum** or valve of Vieussens. The **nodule** is connected to the flocculi by the **inferior medullary velum.** The uvula is connected to the amygdalæ by the **furrowed bands.** (Morris, pp. 736-739; Gray, pp. 724-732.)

Fissures of the cerebellum. The **great horizontal fissure** separates the superior surface from the inferior surface of the cerebellum. On the superior surface are; the **precentral fissure,** separating the lobus lingualis from the lobus centralis; the **postcentral fissure,** separating the lobus centralis from the lobus culminis; the **preclival fissure,** separating the lobus culminis from the lobus clivi; and the **postclival fissure,** separating the lobus clivi from the lobus cacuminis. On the inferior surface are; the **postnodular fissure,** separating the lobus nodulæ from the lobus uvulæ; the **prepyramidal fissure,** separating the lobus uvulæ from the lobus pyramidalis; and the **postpyramidal fissure,** separating the lobus pyramidalis from the lobus tuberis.

In the cerebellum the gray matter is on the surface and the white matter is in the substance of the organ. On section, the appearance of the central core of white matter, sending a branch into each folium of the gray matter, gives rise to the term **arbor vitæ** of the cerebellum.

The cortex of the cerebellum is divisible into an outer, molecular layer, a single layer of the ganglion cells of Purkinje, and an inner, granular layer.

We find five groups of cells in the cerebellar cortex. (1) The cells of Purkinje; (2) cells in the granular layer, the neurits of which extend into the white matter of the cerebellum; (3) cells in the granular layer, the neurits of which extend into the molecular layer; (4) cells of the second type in the granular layer; and (5) the basket cells, which are situated in the molecular layer. The neurits of these cells form basket-like ramifications around the bodies of the cells of Purkinje.

The gray matter in the substance of the cerebellum is collected into four pairs of nuclei; the **nucleus dentatus,** and the three **roof nuclei** on each side. (Morris, p. 739; Gray, p. 735.)

The cerebellum is connected to the remainder of the central nervous system by three pairs of peduncles; the superior, the middle, and the inferior peduncles of the cerebellum. The **superior peduncles** enter the cerebellum from the posterior pair of corpora quadrigemina; the **middle peduncles** enter the cerebellum from the pons; and the **inferior peduncles** enter from the medulla, as the restiform bodies. (Morris, p. 737; Gray, p. 733.)

The fibres which enter the cerebellum from the superior peduncle pass to the dentate nucleus. Of the fibres which enter from the inferior peduncle, some pass through, others pass superficial to, and still others pass deeper than the dentate nucleus. The fibres which enter from the middle peduncles come from the pontine nuclei; those which are situated higher in the pons pass to the inferior portion of the cerebellum; while those which lie lower in the pons pass to the superior portion of the cerebellum. **Association fibres** are those which connect adjacent or neighboring folia of the cerebellum.

THE FOURTH VENTRICLE.

The **fourth ventricle** is a lozenge-shaped space having a floor, a roof, and sides. The roof of the fourth ventricle is formed by the **anterior medullary velum** and the **posterior medullary velum.** The floor of the fourth ventricle is formed by the **pons** and the **medulla.** The sides of the fourth ventricle are formed, anteriorly, by the **superior peduncles of the cerebellum,** and posteriorly, by the **inferior peduncles of the cerebellum.** In the roof of the fourth ventricle we find the **foramen of Majendie,** which opens into the cisterna magna, and the choroid plexus of the fourth ventricle. The inferior medullary velum is separated from the cavity of the fourth ventricle by the ependyma. The **ependyma** is the single layer of ciliated columnar epithelial cells and its basement membrane of neuroglia, which lines the ventricular cavities of the brain and the central canal of the spinal cord.

The floor of the fourth ventricle is bisected longitudinally by the **median fissure,** which terminates, inferiorly, in a peculiar marking known as the **calamus scriptorius,** which resembles the nib of a quill pen. The floor of the fourth ventricle is divided into an anterior half and a posterior half by transverse bundles of fibres, the **striæ acusticæ,** which appear to come from the median fissure and which pass outward over the restiform bodies. The posterior half of the floor of the fourth ventricle presents for study; (1) the **trigonum hypoglossi,** (2) the **trigonum vagi,** and (3) the **tuberculum acusticum** or **auditory tubercle.** The anterior half of the floor of the fourth ventricle is crossed by a band of fibres known as the **conductor sonorus ;** and presents for study; (1) the **eminentia teres,** (2) the **locus ceruleus,** and (3) the **fovea superior.**

The trigonum hypoglossi is formed by the underlying nucleus of origin of the hypoglossal nerve. The trigonum vagi is formed by the underlying nuclei of origin of the pneumogastric and glosso-pharyngeal nerves. The auditory tubercle is formed by the underlying auditory nucleus. The eminentia teres is formed by the underlying fibres of the facial nerve. The locus ceruleus is formed by underlying masses of gray matter. (Morris, pp. 739-741 ; Gray, pp. 723 and 737.)

THE CRURA CEREBRI.

The **crura cerebri** are two large bundles of nerve fibres, which come off from the superior border of the pons and pass into the cerebrum.

The cerebral crus may be divided into an anterior part, or **crusta,** and a posterior part, or **tegmentum.** Between the crusta and the tegmentum is a large collection of gray matter, the **locus niger.**

The **crusta** is composed of longitudinal fibres, which are the continuations of the longitudinal fibres of the pons and of the fibres in the pyramids of the medulla. The tegmentum contains the nucleus rubrum, the body of Luys, the formatio reticularis, the superior or mesial fillet, and the inferior or lateral fillet.

The **superior or mesial fillet** begins in the formatio reticularis of the medulla, passes upward through the medulla, pons, and crus cerebri to terminate in the superior corpus quadrigeminum. It is composed (1) of the continuation upward of the internal arcuate fibres (see page 26), and (2) of fibres from the pontine nuclei.

The **inferior or lateral fillet** begins in the pons Varolii and ends in the inferior corpus quadrigeminum. It is composed; (1) of fibres from the superior olive, (2) of fibres from the superior lateral area of the cerebral crus, (3) of fibres from the pontine nuclei, (4) of fibres from the accessory auditory nucleus, and (5) of fibres from the antero-lateral tract of the spinal cord. (Morris, p. 735; Gray, p. 740.)

THE CEREBRUM.

The **cerebrum** is divided by the **longitudinal fissure** into the right hemisphere and the left hemisphere. The **transverse fissure** is prolonged inward between the splenium of the corpus callosum and the corpora quadrigemina and then downward, just above the corpus fimbriatum. The latter part is called the **inferior fissure.**

Each hemisphere may be studied from its convex surface, from its mesial surface, and from its inferior surface. On each of these surfaces one is able to distinguish certain fissures or sulci and certain convolutions or gyri. The fissures may be classified as fundamental fissures, as interlobar fissures, and as secondary fissures.

A **fundamental fissure** is one which involves the entire thickness of the cerebral cortex so that a corresponding elevation is seen on the underlying ventricular wall.

An **interlobar fissure** is one which serves to divide the cerebral cortex into certain lobes.

A **secondary fissure** is one which divides the lobes of the cerebral cortex into convolutions.

The fundamental fissures are: (1) the *fissure of Sylvius*, (2) the *calcarine fissure*, (3) the *collateral fissure*, and (4) the *dentate* or *hippocampal fissure*.

The interlobar fissures are: (1) The *fissure of Sylvius*, (2) the *fissure of Rolando*, (3) the *parieto-occipital fissure*, (4) the *calloso-marginal fissure*, and (5) the *collateral fissure*.

The **fissure of Sylvius** begins at the anterior perforated space, at the base of the brain, and, passing outward and backward, is seen on the convexity of the hemisphere. It divides into an anterior limb, running forward, an ascending limb, running upward, and a horizontal limb, running backward.

The **fissure of Rolando** begins in the longitudinal fissure a short distance behind the middle point of the cerebral hemisphere. It forms an angle of about 71.7° with the longitudinal fissure and an angle of about 143.4° with its fellow of the opposite side. The fissure passes obliquely downward and forward to end just above the fissure of Sylvius. In its course it presents two bends, the **superior genu** and the **inferior genu.**

The **parieto-occipital fissure** is composed of two parts, the external limb, seen on the convexity of the hemisphere, and the internal limb seen on the mesial surface of the hemisphere. The **internal limb** begins at the isthmus, and, passing upward and backward, is continuous across the margin of the hemisphere with the **external limb.** The external limb is about one-half inch long.

The **calloso-marginal fissure** commences below the anterior extremity of the corpus callosum and passes upward and backward, midway between the corpus callosum and the margin of the hemisphere. About opposite the splenium of the corpus callosum it turns upward and ends at the margin of the hemisphere, just behind the fissure of Rolando. That portion of the calloso-marginal fissure which lies in front of the anterior end of the corpus callosum is called the **prelimbic fissure.**

The **collateral fissure** begins on the posterior margin of the hemisphere in the region of the occipital lobe. It runs forward to end just in front of the uncinate gyrus. (Morris, p. 706; Gray, p. 771.)

These fissures divide the cerebral cortex into six lobes: (1) the *frontal*, (2) the *parietal*, (3) the *occipital*, (4) the *temporal*, (5) the *limbic*, and (6) the *island of Reil* or *central*.

The **frontal lobe** is bounded in front and above by the margin of the hemisphere, below by the fissure of Sylvius, and behind by the fissure of Rolando. It extends on the mesial surface as far as the calloso-marginal fissure. The frontal lobe has an **orbital** or **inferior surface** which rests on the orbital plate of the frontal bone and on the lesser wing of the sphenoid bone. In the frontal lobe may be seen (1) the **precentral fissure,** in front of and parallel to the fissure of Rolando, (2) the **superior,** and (3) the **inferior frontal fissures** running fore and aft, or at right angles to the precentral fissure. These three fissures are to be seen on the convexity of the frontal lobe and divide that lobe into (1) the **ascending frontal convolution,** (2) the **superior frontal convolution,** (3) the **middle frontal convolution,** and (4) the **inferior frontal convolution.** The ascending frontal convolution lies parallel with the fissure of Rolando while the superior, middle, and inferior frontal convolutions have a long axis which is at right angles with that of the ascending frontal convolution. On the mesial surface of the frontal lobe we see the **marginal convolution** lying between the margin of the hemisphere and the calloso-marginal fissure. It may be noticed, also, that a secondary fissure begins in the calloso-marginal fissure just before that fissure turns at right angles to pass up to the margin of the hemisphere. This secondary fissure, known as the **paracentral fissure,** also passes up toward the margin of the hemisphere and between the two fissures a small convolution is included, of which part belongs to the frontal lobe and part to the parietal lobe. From the fact that the fissure of Rolando begins either on the margin of, or just within this convolution it has been termed the **paracentral lobule** or convolution.

The inferior surface of the frontal lobe presents the olfactory fissure and the triradiate fissure. The **olfactory fissure** accommodates the olfactory tract. Between the longitudinal fissure of the brain and the olfactory fissure we find the **gyrus rectus** and between the limbs of the **triradiate fissure** we see the **internal, anterior, and posterior orbital convolutions.** The internal, anterior, and posterior orbital convolutions are continuous, respectively, with the superior, middle, and inferior convolutions as defined on the convexity of the frontal lobe. (Morris, p. 708; Gray, p. 775.)

ANALYSIS OF THE FRONTAL LOBE.

I. Fissures:
  a, on convexity;
    1, precentral,
    2, superior,
    3, inferior;
  b, on mesial surface;
    1, paracentral;
  c, on inferior surface;
    1, olfactory,
    2, triradiate.

II. Convolutions:
  a, on convexity;
    1, ascending frontal,
    2, superior frontal,
    3, middle frontal,
    4, inferior frontal. (Convolution of Broca, See p. 39.)
  b, on mesial surface;
    1, marginal,
    2, part of paracentral;
  c, on inferior surface,
    1, gyrus rectus,
    2, internal orbital,
    3, anterior orbital,
    4, posterior orbital.

The **parietal lobe** is bounded in front by the fissure of Rolando, below by the fissure of Sylvius, and behind by the occipital lobe, into which it blends. On the convexity of the lobe we see the **intraparietal fissure.** This fissure begins a short distance behind the fissure of Rolando and runs parallel with that fissure for about two-thirds of its course, when it bends sharply at right angles, passes backward and ends on the boundary between the occipital and parietal lobes. Occasionally this fissure is continuous with the anterior occipital fissure. From the point at which the intraparietal fissure bends backward, the **superior vertical limb** of the fissure is continued upward, parallel with the fissure of Rolando. This fissure divides the parietal lobe into the **ascending parietal convolution,** between its vertical limbs and the fissure of Rolando; the **superior parietal convolution,** between the horizontal limb and the margin of the hemisphere; and the **inferior parietal convolution,** between the horizontal limb and the fissure of Sylvius. The inferior parietal convolution is subdivided into the **supra-marginal gyrus,** the **angular gyrus,** and the **post-parietal gyrus.** (See p. 39.) On the mesial surface of the parietal lobe we see that the **precentral lobule** contains the mesial portion of the ascending parietal convolution. Between the ascending limb of the calloso-marginal fissure and the internal parieto-occipital fissure we have a four-sided convolution, which is separated from the underlying portion of the limbic lobe by the **postlimbic fissure.** This is the **quadrate lobule** or **precuneus.** (Morris, p. 709; Gray, p. 776.)

ANALYSIS OF THE PARIETAL LOBE.

1. Fissures:
  a, on convexity;
    1, intraparietal;
  b, on mesial surface;
    1, postlimbic.

II. Convolutions:
  a, on convexity;
    1, ascending parietal,
    2, superior parietal,
    3, inferior parietal;
  b, on mesial surface;
    1, part of paracentral,
    2, precuneus or quadrate.

The **occipital lobe** is bounded posteriorly by the margin of the hemisphere and anteriorly by a line drawn from the external parieto-occipital fissure, through the anterior and lateral occipital fissures, to the lower margin of the hemisphere. The anterior portion of the occipital lobe and the posterior portions of the parietal and temporal lobes are very closely related and a sharp line of division is well-nigh impossible. The occipital lobe, on the mesial surface, is separated from the temporal lobe by the **collateral fissure.**

On the convexity of the hemisphere we see the **anterior occipital** and the **lateral occipital fissures.** These fissures lie almost at right angles to each other. They divide the occipital lobe into the **superior, middle,** and **inferior occipital convolutions.** On the mesial surface of the occipital lobe we see the **calcarine fissure,** which begins in a forked extremity near the lower portion of the occipital lobe and which passes upward and inward to end in the internal parieto-occipital fissure. The triangular convolution situated between these two fissures is known as the **cuneus.** Between the calcarine fissure and the collateral fissure we see the **lingual lobule,** part of which belongs to the occipital lobe. (Morris, p. 710; Gray, p. 777.)

ANALYSIS OF THE OCCIPITAL LOBE.

I. Fissures :
  a, on convexity;
    1, anterior,
    2, lateral ;
  b, on mesial surface;
    1, calcarine.

II. Convolutions :
  a, on convexity;
    1, superior,
    2, middle,
    3, inferior ;
  b, on mesial surface; '
    1, cuneus,
    2, part of lingual lobule.

The **temporal lobe** is bounded above by the fissure of Sylvius; below, it continues around the inferior margin of the hemisphere on to the mesial surface to be divided from the limbic lobe, and from the occipital lobe by the collateral fissure. On the convexity of the hemisphere we see the **superior, middle,** and **inferior temporal fissures.** The superior temporal fissure is also called the **parallel fissure.** The parallel and the middle temporal fissures extend backward and upward, parallel to the horizontal limb of the fissure of Sylvius, to end in the inferior parietal convolution. These fissures define the **superior, middle,** and **inferior temporal convolutions.** The inferior temporal fissure is usually seen just at the inferior margin of the hemisphere.

On the mesial surface we see the collateral fissure which divides the fourth temporal or **fusiform convolution** from the limbic lobe. The collateral fissure extends backward into the region of the occipital lobe and between it and the calcarine fissure we have the **lingual convolution**, which belongs partly to the temporal and partly to the occipital lobe. (Morris, p. 712; Gray, p. 777.)

ANALYSIS OF THE TEMPORAL LOBE.

I. Fissures:
    a, on convexity;
        1, parallel (superior),
        2, middle,
        3, inferior.

II. Convolutions:
    a, on convexity;
        1, superior,
        2, middle,
        3, inferior,
    b, on mesial surface;
        1, fusiform,
        2, part of lingual.

The **limbic lobe** is that portion of the mesial surface of the cerebral hemisphere which lies in a concentric manner around the corpus callosum. It is separated from the body of the corpus callosum by the **callosal fissure**. It is separated from the splenium of the corpus callosum by the **dentate fissure.** The dentate and callosal fissures are continuous. The limbic lobe is composed of (1) the *gyrus fornicatus*, (2) the *isthmus*, (3) the *hippocampal gyrus*, and (4) the *uncinate gyrus*. The **gyrus fornicatus**, is separated from the marginal gyrus by the calloso-marginal fissure and from the precuneus by the postlimbic fissure. The **isthmus** is the continuation of the gyrus fornicatus backward between the splenium of the corpus callosum, from which it is separated by the dentate fissure, and the cuneus and lingual convolution, from which it is separated by the internal parieto-occipital fissure. The **hippocampal gyrus** is the continuation forward of the isthmus; it is separated from the dentate fascia by the dentate fissure and from the fusiform convolution by the collateral fissure. The **uncinate gyrus** is the hook-shaped termination of the hippocampal gyrus; it lies beneath the optic thalamus and is limited anteriorly by the beginning of the fissure of Sylvius. The **dentate fascia** is an atrophied portion of the cerebral cortex, situated between the dentate fissure and the optic thalamus. (Morris, p. 717; Gray, p. 781.)

The **island of Reil,** also called the **central lobe,** is found at the bottom of the fissure of Silvius. It is a portion of the cerebral cortex which, from its close attachment to the underlying masses of gray matter, becomes covered up by the expansion of the remainder of the pallium. The island of Reil is separated from the remainder of the cerebral substance by the **limiting fissure** and is divided, by the **central fissure** of the island of Reil, into an anterior, **gyrus longus**, and three or four posterior, **gyri breves.** (Morris, p. 713; Gray, p. 778.)

The **convolution of Broca** is that portion of the inferior frontal convolution which winds around the ends of the anterior and ascending limbs of the fissure of Sylvius. It is the speech centre and is better developed on the left side than on the right.

The **operculum** is the part of the ascending frontal and of the ascending parietal convolutions which meet around the lower end of the fissure of Rolando. It overhangs the island of Reil and, lid-like, occludes the fissure of Sylvius.

The **supramarginal convolution** is that part of the inferior parietal convolution which winds around the end of the horizontal limb of the fissure of Sylvius.

The **angular gyrus** is that part of the inferior parietal convolution which winds around the end of the parallel fissure.

The **post-parietal convolution** is that part of the inferior parietal convolution which winds around the end of the middle temporal fissure.

An **annectant gyrus** is one which connects two convolutions by bridging over a fissure, usually one of the secondary fissures.

The cerebral cortex is divisible into the following layers: (1) the layer of neuroglia, (2) the layer of triangular nerve

cells, (3) the layer of pyramidal ganglion cells, (4) the layer
of polymorphous nerve cells, and (5) the layer of non-medullated nerve fibres. (Piersol, p. 311.)

THE LATERAL VENTRICLES.

The **lateral ventricles** are situated in the substance of
the cerebral hemispheres. Each lateral ventricle is composed of
a *body,* an *anterior horn,* a *posterior horn,* and a *descending
horn.*

The **roof** of the lateral venticle is formed by the *corpus
callosum.* The **floor** of the lateral ventricle is formed by (1)
the *caudate nucleus,* (2) the *tenia semicircularis,* (3) the *optic
thalamus,* (4) the *choroid plexus,* and (5) the *corpus fimbriatum.*
Anteriorly the two lateral ventricles are separated from each
other by the *septum lucidum.*

The **corpus callosum** is seen at the bottom of the longitudinal fissure of the brain. It is composed of a *body,* an
anterior extremity or *rostrum,* and a posterior end or *splenium.*
The **rostrum** is connected to the body of the corpus callosum
by the **genu.** On the superior surface of the corpus callosum
we may see a **median raphe** and the **mesial** and the **lateral longitudinal striæ.** These latter markings represent
atrophied portions of the cerebral cortex. The mesial longitudinal striæ are known as the **nerves of Lancissi.** Micro-
scopically, the corpus callosum is composed of transverse
fibres which connect similar points in the two hemispheres
of the cerebrum and is, therefore, a true commissure. The
fibres contained in the corpus callosum pass to the frontal,
the parietal, and the occipital lobes. The fibres passing from
the corpus callosum into the frontal lobe of the cerebrum
constitute the **forceps minor;** those passing into the occipital
lobe form the **forceps major.** (Morris p. 717; Gray p. 756.)

The **caudate nucleus** is a portion of the striate body. (See
p. 43.)

The **optic thalamus** is one of the basal gray ganglia of
the cerebrum. (See p. 45.)

The **tenia semicircularis** is a bundle of nerve fibres which begins in the anterior pillar of the fornix and which ends in the amygdaloid nucleus. It lies between the caudate nucleus and the optic thalamus. (Morris, p. 725; Gray, p. 760.)

The **choroid plexus** is a plexus of veins which is contained in the free margin of the velum interpositum. (See p. 44.)

The **corpus fimbriatum** is the free margin of the posterior pillars of the fornix.

The **anterior horn** of the lateral ventricle projects into the frontal lobe of the cerebrum.

The **posterior horn** of the lateral ventricle projects into the occipital lobe of the cerebrum. On its wall we see a prominence, formed by the projection inward of the calcarine fissure, which is known as the **hippocampus minor, ergot,** or **calcar avis.** Between the posterior horn and the descending horn of the lateral ventricle we may observe a smooth area, termed the **trigonum ventriculi.** (Morris, p. 720; Gray, p. 758.)

The **descending horn** of the lateral ventricle projects into the temporal lobe of the cerebrum. In it we see (1) the *hippocampus major,* ending in the *pes hippocampi,* (2) the *eminentia collateralis,* (3) the *corpus fimbriatum,* (4) the *dentate fascia,* (5) the *choroid plexus,* (6) the *tenia semicircularis,* (7) the tail of the *caudate nucleus,* and (8) the *amygdaloid nucleus.*

The **hippocampus major** is a bulging on the inner wall of the descending horn of the lateral ventricle, formed by the projection inward of the dentate fissure. It presents, at its anterior extremity, a convoluted mass termed the **pes hippocampi.** (Morris, p. 720; Gray, p. 763.)

The **eminentia collateralis** is formed by the projection inward of the collateral fissure.

The **dentate fascia** is an atrophied portion of the cerebral cortex which is situated between the hippocampus major and the corpus fimbriatum. On the mesial surface of the cerebrum the dentate fascia is seen as a serrated structure between the optic thalamus and the dentate fissure. (Morris, p. 721; Gray, p. 765.)

The **choroid plexus** is a plexus of veins contained in

the free margin of the velum interpositum. In this situation, as well as in the floor of the body of the ventricle, the velum interpositum is separated from the cavity of the ventricle by the ependyma. It occupies the inferior fissure.

The **amygdaloid nucleus** is a thickening of the cerebral cortex found in the anterior extremity of the descending horn of the lateral ventricle. (Morris, p. 725; Gray, p. 760.)

The **septum lucidum** is an atrophied portion of the cerebral cortex, situated beneath the corpus callosum, above and in front of the fornix, and between the lateral ventricles. It contains a small cleft, which is an isolated portion of the longitudinal fissure, formed by the development of the corpus callosum. This cleft is the so-called **fifth ventricle.** It is not a true ventricle. (Morris, p. 726; Gray, p. 762.)

The **fornix** is a longitudinal commissure of white fibres which lies beneath the corpus callosum and above the velum interpositum. It is composed of a *body,* a pair of *anterior pillars*, and a pair of *posterior pillars.* The **anterior pillars** of the fornix form almost a right angle with the body of the fornix and pass down to the base of the brain to terminate in the corpora albicantia. The fibres contained in the anterior pillars of the fornix form terminal arborizations around the cells in the corpora albicantia, and from these cells neurits arise which pass to the optic thalamus as the **bundle of Vicq d' Azyr.** The **posterior pillars** of the fornix are deflected from the median line, and each, by its free margin, projects into the body and the descending horn of the corresponding lateral ventricle. In this situation the posterior pillar of the fornix is known as the **corpus fimbriatum.** The posterior pillars of the fornix end in the amygdaloid nuclei. A mixed transverse and longitudinal striation of the body of the fornix between the two diverging posterior pillars is termed the **lyre.** The fornix and the corpus fimbriatum together are morphologically the white matter corresponding to the dentate fascia and the nerves of Lancissi. Collectively, these structures are known as the **inferior limbic lobe** or the **gyrus dentatus.** (Morris, p. 725; Gray, p. 760.)

The **striate body** is a collection of gray matter situated at the base of the brain. It is composed of two parts; the *caudate nucleus* and the *lenticular nucleus*. From the fact that the caudate nucleus lies nearer the midline, and forms a projection on the floor of the lateral ventricle, it is termed the **intraventricular portion** of the striate body. On the other hand, the lenticular nucleus is known as the **extraventricular portion** of the striate body, because it does not come in relation with the lateral ventricle. The caudate and lenticular nuclei are continuous with each other anteriorly; but, in the posterior part of their course, they are separated from each other by the internal capsule.

The **caudate nucleus** is a collection of gray matter, resembling a crook-necked squash in shape. Its **head,** or large, anterior extremity, is seen in the floor of the lateral ventricle, while its **tail,** or narrow, posterior extremity is found in the roof of the descending horn of the lateral ventricle.

RELATIONS.—Anteriorly, it forms the posterior wall of the anterior horn of the lateral ventricle, and is continuous with the lenticular nucleus. Posteriorly, it is separated from the optic thalamus by the tenia semicircularis, whilst its tail is seen in the roof of the descending horn of the lateral ventricle. Internally, it is in relation with the septum lucidum. Superiorly, it is in relation with the lateral ventricle. Externally and inferiorly, it is separated from the lenticular nucleus by the internal capsule.

The **lenticular nucleus** is a pyramidal mass of gray matter, composed of an outer portion or **putamen** and an inner segment or **globus pallidus.**

RELATIONS.—Superiorly it is in relation with the substance of the cerebral hemisphere. Inferiorly, it is in relation with a portion of the anterior perforated space. Internally it is in relation with the internal capsule, which separates it from the caudate nucleus in front and the optic thalamus behind. Externally, it is in relation with the external capsule, which separates it from the claustrum. (Morris, p. 722; Gray. p. 759.)

The **internal capsule** is a longitudinal band of white

fibres which connects the motor region of the cerebral cortex, around the fissure of Rolando, with the motor tracts
below. In its course it lies; first, between the lenticular and
caudate nuclei; and second, between the lenticular nucleus and
the optic thalamus. Its fibres are then continued through
the crusta of the cerebral crus, the crusta of the pons, and
the pyramids of the medulla, into the anterior and lateral
pyramidal tracts of the spinal cord. The **corona radiata** is
the structure formed by the converging of the fibres from the
cerebral cortex to enter the internal capsule. (Morris, p. 725;
Gray, p. 760.)

The **external capsule** is a band of white fibres which
separates the lenticular nucleus from the claustrum.

The **claustrum** is a small, isolated band of gray matter
lying between the external capsule and the white matter of
the island of Reil. (Morris, p. 724; Gray, p. 760.)

If a knife is pushed into the cerebrum from the bottom
of the fissure of Sylvius it will cut the following structures:
(1) the island of Reil, (2) the claustrum, (3) the external
capsule, (4) the lenticular nucleus, and (5) the internal capsule. If, now, the knife is directed slightly backward it will
pass through: (6) the optic thalamus, into (7) the third ventricle. If, on the other hand, the knife is directed somewhat forward and upward it will cut: (6) the caudate
nucleus, and enter (7) the body of the lateral ventricle. (See
diagram Morris, p. 724; Gray, p. 758.)

THE THIRD VENTRICLE.

The **roof** of the third ventricle is formed by the *velum
interpositum*. The **sides** of the third ventricle are formed by
the *optic thalami*. The **floor** of the third ventricle is formed
by (1) the *lamina cinerea*, (2) the *optic commissure*, (3) the
anterior perforated spaces, (4) the *tuber cinereum*, (5) the
corpora albicantia or *mamillaria*, (6) the *in fundibulum*, (7)
the *pituitary body*, and (8) the *posterior perforated space*.

The **velum interpositum** is a process of pia mater

which grows in at the transverse fissure of the cerebrum from before backward and which lies, just beneath the fornix. It is separated from the cavity of the ventricle by the ependyma. The velum interpositum projects by its free margin into the floor and descending horn of each lateral ventricle. As it passes across the superior surface of the optic thalamus it is attached to that body along the **sulcus choroideus.** It contains the *veins of Galen* and the *choroid plexuses* of the lateral and third ventricles. (Morris, p. 727; Gray, p. 749.)

The **optic thalamus** is a mass of gray matter which lies in relation partly with the third ventricle and partly with the floor of the lateral ventricle. It is composed of an **anterior tubercle** and a **posterior tubercle** or **pulvinar.** In this structure the gray matter is collected to form an **anterior nucleus,** a **mesial nucleus,** and a **lateral nucleus.** These nuclei are connected to each other by numerous bands of medullated nerve fibres, which are known as the **medullary striæ.** The gray matter of the optic thalamus is separated from the ependyma by a distinct layer of white matter, which is known as the **stratum zonale.** A bundle of white fibres which passes from the optic thalamus to the lenticular nucleus, lying beneath the internal capsule in its course, is known as the **ansa lenticularis.** The **inferior peduncle of the optic thalamus** is a bundle of nerve fibres which passes from the pulvinar into the internal capsule. On the border between the superior and internal surfaces of the optic thalamus a distinct band of fibres is to be seen which is the **peduncle of the pineal body.** Between this band of fibres and the superior surface of the optic thalamus there is a triangular depressed area which is known as the **trigonum habenulæ.** The **sulcus choroideus** is a groove on the superior surface of the optic thalamus along which the velum interpositum is attached.

RELATIONS.—Superiorly, the optic thalamus is in relation with the velum interpositum, which is attached to it along the sulcus choroideus, and with the cavity of the body of the lateral ventricle. Anteriorly, the optic thalamus is separated from the caudate nucleus by the tenia semicircularis and forms the posterior boundary of the foramen of Munro. Internally, the optic thalamus is in relation with the cavity of the third ventricle. Posteriorly, the optic thalamus overhangs the geniculate bodies and is in relation with the anterior pair of corpora quadrigemina. Inferiorly the optic thalamus is in relation with the subthalamic region. Externally, the optic thalamus is separated from the lenticular nucleus by the internal capsule. (Morris, p. 729; Gray, p. 746.)

The **lamina cinerea** and the **tuber cinereum** are thin masses of gray matter which help to form the floor of the third ventricle.

The **anterior perforated spaces** are two thin layers of gray matter which contain numerous small foramina for the passage of blood vessels.

The **corpora albicantia** or **mamillaria** are two knob-like masses seen in the floor of the third ventricle. Each body contains a nucleus of gray matter, in which the anterior pillar of the fornix ends and the bundle of Vicq d' Azyr begins.

The **in fundibulum** is a small diverticulum from the floor of the third ventricle, which bears the pituitary body. It contains a narrow cleft, which is the continuation of the cavity of the ventricle.

The **pituitary body** is a small, rounded structure, which is composed of an anterior and a posterior lobe. It is contained in the sella turcica. It has, possibly, something to do with the growth of the body.

The **posterior perforated space** is a thin layer of gray matter which is seen in the floor of the third ventricle. It contains numerous small foramina for the passage of blood vessels. It is situated in the angle formed by the diverging crura cerebri. (Morris, p. 732; Gray, pp. 748-751.)

The **subthalamic region** is that portion of the cerebrum which lies between the optic thalami and the tegmenta of the crura cerebri. This region contains the *body of Luys,* the *nucleus rubrum,* the *ansa lenticularis,* and the *inferior peduncle of the optic thalamus;* these structures are paired; one of a pair being found to the right and the other to the left of the mid-line. (Morris, p. 736; Gray, p. 745.)

The **cavity** of the third ventricle is composed of a vertical portion, which lies between the two optic thalami, and a horizontal portion, which lies between the velum interpositum and the upper surface of the optic thalamus, on either side.

Three commissures cross the vertical portion of the ventricle: the anterior and posterior, white, and the middle, gray commissures. The **anterior commissure** is a true commissure which connects similar parts in the two temporal lobes of the cerebrum. It lies below the anterior pillars of the fornix, and supplements the corpus callosum.

The **middle commissure** connects the two optic thalami.

The **posterior commissure** contains fibres which are decussating. It is not a true commissure. (Morris, pp. 728 and 731 ; Gray, p. 745.)

The **pineal body** is situated in the roof of the third ventricle. It is connected to the optic thalami by the **peduncles of the pineal body** and with the tissue below the corpora quadrigemina by a bundle of fibres, which, together with the peduncles, forms the **stalk** of the pineal body. By its inferior surface the pineal body rests in the groove between the anterior pair of corpora quadrigemina. It is a rudimentary sense organ. (Morris, p. 730; Gray, p. 748.)

The **foramen of Munro** is a Y-shaped space which opens by its stem into the third ventricle and by its two arms into the two lateral ventricles. It is bounded in front by the anterior pillar of the fornix and behind by the optic thalamus. (Morris, p. 721; Gray, p. 748.)

THE AQUEDUCT OF SYLVIUS.

The **aqueduct of Sylvius** is the passage from the third into the fourth ventricle.

The **roof** of the aqueduct of Sylvius is formed by the *lamina quadrigemina*. The **floor** of the aqueduct of Sylvius is formed by the *crura cerebri*.

48 THE CENTRAL NERVOUS SYSTEM.

The **corpora quadrigemina** are four small bodies known as an anterior pair or **nates** and a posterior pair or **testes.** The corpora quadrigemina are separated from the aqueduct of Sylvius by a plate of white matter, known as the **lamina quadrigemina.** The anterior pair of corpora quadrigemina are connected, by the **superior brachia,** with the right and left *external geniculate bodies.* The posterior pair of corpora quadrigemina are connected, by the **inferior brachia,** with the right and left *internal geniculate bodies.* In order to see the geniculate bodies it is necessary to pull up the pulvinar of the optic thalamus, beneath which they lie. (Morris, p. 733; Gray, p. 743.)

THE ASSOCIATION FIBRES OF THE CEREBRUM.

The different convolutions of the cerebrum are connected with each other by tracts of fibres which are known as **association fibres.** These association fibres may be divided into a short group and a long group. The **short** association fibres pass between adjacent or neighboring convolutions of the cerebral cortex. The **long** association fibres may be divided as follows: (1) the **superior longitudinal fasciculus,** which connects the frontal lobe with the occipital lobe. (2) The **inferior longitudinal fasciculus,** which connects the occipital lobe with the anterior portion of the temporal lobe. (3) The **uncinate fasciculus,** which connects the inferior portion of the frontal lobe with the uncinate gyrus. (4) The **posterior fasciculus,** which connects the temporal lobe with the parietal lobe. (5) The **cingulum,** which lies in the limbic lobe, connecting the two extremeties of that lobe. (6) The **corpus fimbriatum,** the **fornix,** and the **bundle of Vicq d' Azyr.** (Piersol, p. 326.)

THE DEVELOPMENT OF THE NERVOUS SYSTEM.

The nervous system is developed from the walls of the neural canal, which, in turn, are formed by the union of the medullary folds of the ectoderm. The cells in the walls of the neural canal are of two kinds; the *spongioblasts* and the *neuroblasts.*

The **neuroblasts** form the nerve cells of the brain and spinal cord. The **spongioblasts** form the neuroglia and the cells of the ependyma.

The **spinal cord** is formed by a thickening of the wall of the neural canal. The canal itself, which becomes narrower as the wall grows thicker, remains as the **central canal of the spinal cord.**

A small thickening of the dorsal portion of the wall of the neural canal becomes isolated to form the various **ganglia** found throughout the body.

At the cephalic end of the embryo, the neural canal becomes flexed and, by a process of unequal growth, divided into three **primary cerebral vesicles**; the *fore-brain,* the *mid-brain,* and the *hind-brain.* The fore-brain and the hind-brain subsequently become subdivided. As the result of this subdivision five **secondary cerebral vesicles** are formed. The secondary cerebral vesicles are named: the *fore-brain,* the *inter-brain,* the *mid-brain,* the *hind-brain,* and the *after-brain.*

The **fore-brain** becomes divided into two parts by the downgrowth of the primitive falx from the overlying mesoderm. This division results in the formation of the **longitudinal fissure.** The **corpus callosum** grows across the longitudinal fissure, thus isolating a portion of the fissure, which is found beneath the corpus callosum as the **fifth ventricle.** The **nerves of Lancissi** are the remains of the cerebral cortex which was broken through by the growth of the corpus callosum. The **septum lucidum** is formed by the remains of the cerebral cortex below the corpus callosum. The **fornix** is formed by the fusion of the white matter of the two sides of the divided fore-brain. The roof of the fore-brain develops into the **cerebral hemispheres** or **pallium.** The cavities of the divided fore-brain become the **lateral ventricles.**

The **inter-brain.** The **pineal body** develops in the roof of the inter-brain. In its early stage it resembles a fetal eye. It undergoes calcareous change in the adult. The **corpora albicnatia** are developed from the floor of the inter-brain. The **pituitary body** is developed partly from a downgrowth from the floor of the inter-brain and partly from an upgrowth from the primitive pharynx. The roof of the inter-brain is always very thin. The **velum interpositum** is a process of pia mater which grows in at the transverse fissure from before backward, but is separated from the ventricle by the very thin roof, which, in this situation, consists only of the ependyma. The **optic thalamus** is developed from the sides of the inter-brain. The cavity of the inter-brain remains as the **third ventricle.**

The **mid-brain.** The **corpora quadrigemina** develop from the roof of the mid-brain. The **crura cerebri** are formed from the floor of the mid-brain. The cavity of the mid-brain forms the **aqueduct of Sylvius.**

The **hind-brain.** The **cerebellum** and **anterior medullary velum** are developed from the roof of the hind-brain. The **pons Varolii** is formed from the floor of the hind–brain. The cavity of the hind-brain forms part of the **fourth ventricle**.

The **after-brain.** The floor of the after-brain gives us the **medulla oblongata.** The roof is very thin, giving us the **posterior medullary velum,** which is always separated from the cavity by the ependyma, the morphological covering of the vesicle. The cavity of the after-brain enters into the formation of the **fourth ventricle.**

The **transverse fissure** is formed by the growth of the pallium backward, covering in structures which, in the lower types, are plainly on the superior surface of the cerebrum. The **fissure of Sylvius** is formed from the unequal growth of the cerebral cortex which is not anchored to the underlying gray matter. The **island of Reil,** which was originally on the surface of the cerebrum, lies at the bottom of the fissure of Sylvius on account of its relation to the underlying striate body.

The **nerves,** both spinal and cranial, are direct outgrowths from the neuroblasts. The axis cylinder of a nerve fibre is the neurit of a nerve cell. The medullated nerve fibres receive their coating of the white substance of Schwann in the direction in which they convey impulses. (Quain, p. 57; A. T. O., p. 125.)

PRIMARY CEREBRAL VESICLE.	SECONDARY CEREBRAL VESICLE.	ROOF.	FLOOR.	SIDES.	CAVITY.
Fore-brain.	Fore-brain.	Cerebral Hemispheres, Corpus Callosum, Anterior Commissure, Fornix, Septum Lucidum.	Olfactory Lobes, Anterior Perforated Spaces, Nucleus Caudatus, Nucleus Lenticularis.	Same as Roof.	Lateral Ventricle.
	Inter-brain.	Posterior Commissure, Pineal Body.	Optic Chiasm, Tuber cinereum, Infundibulum, Part of Pituitary Body, Corpora Albicantia.	Optic Thalamus.	Third Ventricle.
Mid-brain.	Mid-brain.	Corpora Quadrigemina.	Crura Cerebri, Posterior Perforated Space.	Geniculate Bodies and Brachia.	Aqueduct of Sylvius.
Hind-brain.	Hind-brain.	Cerebellum, Anterior Medullary Velum.	Pons Varolii.	Anterior and Middle Cerebellar Peduncles.	Fourth Ventricle.
	After-brain.	Posterior Medullary Velum.	Medulla Oblongata.	Inferior Cerebellar Peduncles.	

CHAPTER III.

THE CRANIAL NERVES.

The twelve pairs of nerves which arise from the brain are known as the **cranial nerves,** in contradistinction to those nerves which arise from the spinal cord, and which are known as the spinal nerves.

The cranial nerves are: (1) the olfactory, (2) the optic, (3) the oculo-motor, (4) the trochlear, (5) the trifacial, (6) the abducens, (7) the facial, (8) the auditory, (9) the glosso-pharyngeal, (10) the pneumogastric, (11) the spinal accessory, and (12) the hypoglossal.

The cranial nerves from the third to the twelfth, inclusive, arise from a tract of gray matter situated in the floor of the aqueduct of Sylvius and in the floor of the fourth ventricle. Each pair of cranial nerves may be said to have a point of superficial origin as well as a nucleus of deep origin. The **superficial origin** of a cranial nerve is the point at which it is first seen on the surface of the brain. The **nucleus of deep origin** of a cranial nerve is the collection of gray matter in which the cells are situated, which send off the neurits which constitute the axis cylinders of the nerve fibres composing the nerves.

THE OLFACTORY NERVE.

The olfactory nerve apparatus is composed of (1) the *olfactory tract,* (2) the *olfactory bulb,* and (3) the *olfactory nerves.*

The **olfactory tract** has its **superficial origin** by two roots, an external or lateral root, and an internal or mesial root. The **external root** arises from the anterior portion of the hippocampal gyrus and the amygdaloid nucleus. The **internal root** is connected with the anterior extremity of the gyrus fornicatus. The **deep origin** of the olfactory nerve is in the cortex of the olfactory bulb. The **cortical area of smell** is in the limbic lobe.

52

The triangular area between the two diverging roots of the olfactory tract is known as the **trigonum olfactorium.**

Between the mesial root of the olfactory tract and the longitudinal fissure of the cerebrum there is a rounded elevation known as the **area of Broca.**

The olfactory tract lies in the olfactory fissure on the inferior surface of the frontal lobe of the cerebrum, and, by its ventral surface, rests in a groove on the body of the sphenoid bone.

The **olfactory bulb** is an expansion of the olfactory tract which rests on the cribriform plate of the ethmoid bone. The olfactory tract and bulb, morphologically, represent the undeveloped olfactory lobe of the cerebrum, which is seen in those lower animals that possess a high development of the sense of smell. The human olfactory tract contains; (1) an upper, *dorsal layer of gray matter*, (2) a central core of *neuroglia*, which represents the obliterated ventricular space, surrounded by (3) a layer of *white matter*, and (4) a *ventral layer of gray matter.*

The olfactory bulb consists of the following layers: (1) a core of *neuroglia*, (2) a ring of *medullated nerve fibres* surrounding the core of neuroglia, (3) the *stratum granulosum*, (4) the layer of *olfactory glomeruli*, and (5) the layer of *olfactory nerve fibres.*

The **stratum granulosum** contains; (1) small irregular cells and (2) large pyramidal cells. The dendrits of the pyramidal cells may be classed as apical and as lateral processes. The lateral processes form terminal arborizations in the granular layer. The neurits of the pyramidal cells pass toward the dorsal aspect of the bulb and enter the layer of medullated nerve fibres. Some of these fibres become interrupted in the cells in the cortex of the olfactory tract, and others pass through the white matter of the olfactory tract to the temporal lobe of the brain.

The **olfactory glomeruli** are composed of the terminal arborizations of the apical processes of the pyramidal cells found. in the stratum granulosum, and of the olfactory nerve filaments.

The **olfactory nerves** proper are the twenty or thirty bundles of non-medullated nerve fibres which arise from the olfactory cells in the mucous membrane of the nose, and which,

passing through the foramina in the cribriform plate of the ethmoid bone, terminate in the olfactory glomeruli.

In the nose these nerve fibres are distributed to the superior meatus only, by an **external group** and an **internal
group.**

The epithelium of the olfactory portion of the Schneiderian
mucous membrane is composed of the olfactory cells and the
sustentacular cells.   The olfactory cells are rod-shaped neuro-
epithelial cells, the fibres from which constitute the true olfactory
nerves.   (Piersol, p. 323; Morris, p. 770; Gray, p. 792.)

THE OPTIC NERVE.

The optic nerve apparatus consists of (1) the *optic tracts*, (2)
the *optic commissure*, and (3) the *optic nerves*. The **superficial
origin** of the optic tract is from the outer side of the crus
cerebri.   The **deep origin** of the optic tract is from the optic
thalamus, the anterior corpus quadrigeminum, and the external
geniculate body.   The **cortical area of vision** is in the
cuneus.

From the deep origin of the optic tract some fibres pass to
the cortical area of vision and others pass to the nuclei of
origin of the nerves which serve to move the eyeball.   There
are some fibres in the optic tract which pass to the cortical
area of vision directly, without undergoing interruption in the
nucleus of origin of the tract.

The optic tract passes obliquely forward and inward, on
the under surface of the crus cerebri, and unites with the optic
tract of the opposite side, at the anterior portion of the inter-
peduncular space, to form the optic commissure.

The **optic commissure** rests in a groove on the superior
surface of the body of the sphenoid bone.

There are three groups of **fibres in the optic commissure;** one group passes from the optic nerve of one side into
the optic tract of the opposite side; a second group passes
from the optic nerve of one side into the optic tract of the

same side; and the third group passes from the one optic tract into the opposite optic tract.

After the optic tract has passed behind the crus cerebri it is divided into an **internal portion** and an **external portion** by a well-marked furrow. The external portion is composed of the true optic fibres which have their deep origins as detailed above. The internal portion is associated with the internal geniculate body and the posterior corpus quadrigeminum; and the fibres are continued through the optic commissure as the group passing from one optic tract into the opposite optic tract. These fibres constitute the **commissures of Meynert and of Gudden** and are not connected with vision.

The **optic nerves** are given off from the anterior surface of the optic commissure and leave the cranial cavity by passing through the optic foramen. After leaving the optic foramen the optic nerve enters the orbit and pierces the sclerotic and choroid coats of the eyeball, to be distributed to the retina. In its course the optic nerve is enclosed in a sheath formed by the dura mater and the arachnoid. The optic nerves proper are those fibres which extend from the rods and cones to the cells in the cerebral layer of the retina. (See Anatomy of Eye.)

RELATIONS.—As the optic nerve passes through the optic foramen, the ophthalmic artery lies to its outer side and a little below it. Just after the nerve enters the orbit it is surrounded by the tendons of origin of the four recti muscles. Near the sclerotic coat of the eyeball the optic nerve is surrounded by the long and short ciliary arteries and by the ciliary nerves. The ophthalmic ganglion lies just external to it and the arteria centralis retinæ pierces its inferior surface. (Morris, p. 722; Gray, p. 793.)

THE OCULO-MOTOR NERVE.

The **superficial origin** of the oculo-motor nerve is from the inner side of the crus cerebri.

The **deep origin** of the oculo-motor nerve is from a nucleus in the anterior part of the floor of the aqueduct of Sylvius.

The oculo-motor nerve pierces the dura mater just behind the posterior clinoid process of the sphenoid bone, passes through the wall of the cavernous sinus, and divides into a superior division and an inferior division. The two divisions enter the orbit by passing through the sphenoidal fissure between the two heads of the external rectus muscle.

The **superior division** supplies the superior rectus and the levator palpebræ superioris muscles.

The **inferior division** supplies the internal rectus, the inferior rectus, and the inferior oblique muscles.

The nerve to the inferior oblique muscle gives off the motor root to the ophthalmic ganglion. (Morris, p. 774; Gray, p. 794.)

THE TROCHLEAR NERVE.

The **superficial origin** of the trochlear nerve is from the anterior medullary velum (valve of Vieussens).

The **deep origin** of the trochlear nerve is from a nucleus in the posterior part of the floor of the aqueduct of Sylvius.

The trochlear nerve leaves the skull and enters the orbit by passing through the sphenoidal fissure. It supplies the superior oblique muscle. (Morris, p. 775; Gray, p. 796.)

THE TRIFACIAL NERVE.

The **superficial origin** of the trifacial nerve is by a motor root and sensory root from the side of the pons Varolii.

The **deep origin** of the trifacial nerve is from a motor nucleus and a sensory nucleus in the floor of the fourth ventricle. The motor nucleus is assisted by a descending root which comes from the floor of the aqueduct of Sylvius. The sensory nucleus is assisted by an ascending root which comes from as low down in the spinal cord as the origin of the second cervical nerve. Both roots contain crossed fibres which come from nuclei on the side opposite to that from which the larger number of fibres originate.

The two roots pass forward and pierce the dura mater opposite the apex of the petrous portion of the temporal bone.

After it pierces the dura mater the sensory root presents a
large triangular ganglion known as the **Gasserian ganglion.**
The motor root lies between the Gasserian ganglion and the
bone.

From the anterior border of the Gasserian ganglion three
branches are given off which are known, respectively, as the
ophthalmic division, the *superior maxillary division,* and the *inferior maxillary or mandibular division* of the trifacial nerve.
These three branches are purely sensory in function.

THE OPHTHALMIC DIVISION OF THE TRIFACIAL NERVE.

The **ophthalmic division** of the trifacial nerve passes forward from the Gasserian ganglion, through the outer wall of the
cavernous sinus, and then divides into the lachrymal, the frontal,
and the nasal nerves.

The ophthalmic division of the trifacial nerve is the lowest
of the three nerves which pass through the outer wall of the
cavernous sinus; the trochlear nerve lies immediately above it;
and the oculo-motor nerve lies just above the trochlear nerve.
The three branches of the ophthalmic division of the trifacial
nerve present the following **relations** as they pass through the
sphenoidal fissure to enter the orbit: on the superior margin of
the sphenoidal fissure the trochlear nerve is the most internal
structure, the frontal nerve has the middle position, and the lachrymal nerve is the most external structure. On the outer margin
of the fissure the superior division of the oculo-motor nerve is
the highest structure, the nasal nerve lies next below, the inferior division of the oculo-motor nerve lies beneath the latter,
and the abducens nerve is the lowest structure. The ophthalmic
vein lies in the angle between the inner and outer margins of
the fissure.

The **lachrymal nerve** enters the orbit by passing through
the sphenoidal fissure. It then runs along the outer wall of the
orbit to supply the lachrymal gland. The terminal filaments of
the lachrymal nerve pierce the lachrymal gland and the upper

eyelid to supply the skin, the superficial fascia, and the conjunctiva of the upper lid.

The **frontal nerve** enters the orbit by passing through the sphenoidal fissure. It then passes forward above the levator palpebræ superioris muscle and divides into the *supraorbital* nerve and the *supratrochlear* nerve. The **supraorbital nerve** leaves the orbit by passing, in company with the supraorbital artery, through the supraorbital foramen. It is distributed to the skin and superficial fascia of the scalp as far back as the lambda. The **supratrochlear nerve,** after forming a loop of communication with the infratrochlear branch of the nasal nerve, leaves the orbit by passing above the pulley of the superior oblique muscle and over the internal angular process of the frontal bone. It supplies the skin and conjunctiva at the inner canthus of the eye; and the skin and superficial fascia of the scalp in the region of the glabella.

The **nasal nerve** enters the orbit by passing through the sphenoidal fissure and between the two heads of the external rectus muscle; it then passes across the orbit and enters the cranial cavity by passing through the anterior ethmoidal foramen; thence it crosses the cribriform plate of the ethmoid bone, passes through a slit by the side of the crista galli, and enters the nose, where it lies in a groove on the inferior surface of the nasal bone. In its course, the nasal nerve gives off the *sensory root to the ophthalmic ganglion,* the *infratrochlear nerve,* the *two long ciliary nerves,* the *internal nasal branch,* the *external nasal branch,* and the *anterior branch.*

The **infratrochlear nerve** forms a loop of communication with the supratrochlear branch of the frontal nerve and is distributed to the tissues around the inner canthus of the eye and to the skin over the bridge of the nose.

The two **long ciliary nerves** pierce the sclerotic coat of the eye and are distributed to the ciliary region, the iris, and the cornea.

The **internal nasal branch** is distributed to the mucous membrane covering the upper and anterior part of the septum of the nose.

The **external nasal branch** is distributed to the mucous membrane covering the outer wall of the nose.

The **anterior branch** passes between the nasal bone and the latéral cartilage of the nose and is distributed to the skin of the tip of the nose. (Morris, p. 777; Gray, p. 797.)

THE OPHTHALMIC GANGLION.

In the course of the ophthalmic division of the trifacial nerve there is a small ganglion which is called the **ophthalmic, lenticular, or ciliary ganglion.** This body lies external to the optic nerve. The **motor root** of the ophthalmic ganglion comes from the oculo-motor nerve; the **sensory root** comes from the nasal nerve; and the **sympathetic root** comes from the cavernous plexus of the *sympathetic* system.

The **branches** of the *ophthalmic* ganglion are about six or eight in number and are known as the *short ciliary nerves.* The **short ciliary nerves** pierce the sclerotic coat of the eye and are distributed to the ciliary region, the iris, and the cornea. (Morris, p. 779; Gray, p. 799.)

THE SUPERIOR MAXILLARY DIVISION OF THE TRIFACIAL NERVE.

The **superior maxillary division** of the trifacial nerve is a branch of the Gasserian ganglion. It leaves the cranial cavity and enters the spheno-maxillary fossa by passing through the foramen rotundum. It passes across the spheno-maxillary fossa, through the spheno-maxillary fissure, and enters the orbit. It passes, for a short distance, along the floor of the orbit, enters the infraorbital canal, and, passing through the infraorbital foramen, makes its appearance on the face, beneath the levator labii superioris muscle, where it divides into its terminal branches.

In its course the superior maxillary division of the trifacial nerve gives off the following branches: (1) the *meningeal,* (2) the *spheno-palatine,* (3) the *orbital or temporo-malar,* (4) the *posterior dental,* (5) the *middle dental,* (6) the *anterior dental,* (7) the *labial,* (8) the *nasal,* and (9) the *palpebral.*

The **meningeal branches** are distributed to the dura mater.

The **spheno - palatine branches,** two in number, go to Meckel's ganglion as its sensory roots.

The **orbital** or **temporo-malar nerve** is given off just as the superior maxillary nerve is entering the orbit. It then divides into the **temporal branch,** which passes through the spheno-malar foramen, in the suture between the malar and the sphenoid bones, to be distributed to the skin in the temporal region; and the **malar branch,** which passes through the malar foramen, in the malar bone, to be distributed to the skin covering that bone.

The **posterior dental nerves** supply the superior molar teeth and give **gingival branches** to the gums.

The **middle dental nerves** supply the bicuspid teeth.

The **anterior dental nerves** supply the canine and incisor teeth. These nerves frequently pass through the mucous membrane which lines the antrum of Highmore.

The **labial nerves** are distributed to the skin of the upper lip; the **nasal nerves** are distributed to the skin covering the ala of the nose; and the **palpebral nerves** are distributed to the skin of the lower eyelid.

That portion of the superior maxillary division of the trifacial nerve which passes through the infraorbital canal and foramen is known as the **infraorbital nerve.** The terminal branches of the nerve unite with the infraorbital branch of the facial nerve, beneath the levator labii superioris muscle, to form the **infraorbital plexus.** (Morris, p. 780; Gray, p. 801.)

THE SPHENO-MAXILLARY OR MECKEL'S GANGLION.

In the spheno-maxillary fossa, just beneath the superior maxillary division of the trifacial nerve, we find **Meckel's ganglion.**

The **motor root** of Meckel's ganglion comes from the facial nerve, as the great superficial petrosal nerve. The **sensory roots** come from the superior maxillary division of the trifacial nerve, as the spheno-maxillary branches. The **symphatic root** comes from the carotid plexus of the sympathetic system, as the great deep petrosal nerve.

The great superficial petrosal nerve and the great deep petrosal nerve unite in the substance of the cartilage which fills in the middle lacerated foramen, to form the **Vidian nerve.** The Vidian nerve, accompanied by the Vidian artery, passes through the Vidian canal at the base of the pterygoid process of the sphenoid bone, and enters the spheno-maxillary fossa. It then enters the posterior border of Meckel's ganglion.

The branches of Meckel's ganglion are; (1) an *ascending group*, (2) an *anterior group*, (3) a *posterior group*, and (4) a *descending group*.

The **ascending branches** are distributed to the posterior ethmoidal cells and to the sphenoidal cells.

The **anterior branches** are distributed to the mucous membrane covering the superior and middle turbinated bones, and to the mucous membrane covering the septum of the nose. One of the branches which supply the nasal septum is much larger than the others, and is known as the *naso-palatine nerve*.

The **naso-palatine nerve** enters the nose by passing through the spheno-palatine foramen; it then lies in a groove on the vomer. It passes through the foramen of Scarpa, in the anterior palatine fossa, and, in the mucous membrane covering the roof of the mouth, it joins with the anterior palatine nerve to form a plexus.

The **posterior branch** is called the **pterygo-palatine nerve.** It passes through the pterygo-palatine canal, in company with the pterygo-palatine artery, and is distributed to the mucous membrane lining the posterior wall of the pharynx, particularly around the orifice of the Eustachian tube.

The **descending branches** are three in number: the *anterior palatine nerve*, the *posterior palatine nerve*, and the *external palatine nerve*.

The **anterior palatine nerve** passes through the posterior palatine canal, accompanied by the descending palatine artery,

and makes its appearance on the roof of the mouth just behind the last molar tooth.   It then passes forward in a groove
on the hard palate, supplying the mucous membrane of the
roof of the mouth in its course, and forms a plexus with the
termination of the naso-palatine nerve.

The **posterior palatine nerve** passes through the posterior palatine canal and is distributed to the soft palate.

The **external palatine nerve** passes through the accessory
palatine canal and is distributed to the tonsil and the adjacent
mucous membrane.   (Morris, p. 782; Gray, p. 803.)

THE INFERIOR MAXILLARY DIVISION OF THE TRIFACIAL NERVE.

The **inferior maxillary division** of the trifacial nerve is
a branch of the Gasserian ganglion; it leaves the cranial cavity
by passing through the foramen ovale.   In its passage through
the foramen ovale it is accompanied by the motor root of the
trifacial nerve.   As soon as these two nerves enter the zygomatic fossa they unite to form a common trunk, which then
divides into an anterior division and a posterior division.

The branches of the inferior maxillary division of the trifacial nerve are; (a) from the common trunk, (1) the *meningeal,*
(2) the *nerve to internal pterygoid muscle;* (b) from the anterior
division, (3) the *deep temporal,* (4) the *nerve to the masseter muscle,*
(5) the *nerve to the external pterygoid muscle,* (6) the *buccal;* (c)
from the posterior division, (7) the *auriculo-temporal,* (8) the
lingual, and (9) the *inferior dental.*

The **meningeal nerve** passes backward, through the foramen ovale, to supply the dura mater.

The **nerve to the internal pterygoid muscle,** the **deep
temporal nerves,** the **nerve to the masseter muscle,** and
the **nerve to the external pterygoid muscle** supply the
muscles indicated by their names.

The **buccal nerve** is distributed to the mucous membrane
lining the cheek.

The **auriculo-temporal nerve** arises by two roots, which
surround the middle meningeal artery.   The nerve then passes

upward beneath the parotid gland and above the zygoma, in
company with the superficial temporal artery.   It is distributed
to the articulation of the lower jaw, the external auditory meatus,
the parotid gland, the auricle, and to the skin in the temporal
region.   In its course it receives twigs of communication from
the otic ganglion.

The **lingual nerve** passes forward to be distributed to the
mucous membrane covering the tongue.   In its course it lies
in front of the hyo-glossus muscle and is joined by the chorda
tympani nerve, a branch of the facial.   The lingual nerve is a
nerve of common sensation.   The associated fibres of the chorda
tympani nerve are nerves of the special sense of taste.

The **inferior dental nerve** enters the inferior dental canal
by passing, in company with the inferior dental artery, through
the inferior dental foramen.   It gives off the *mylo-hyoid nerve*
just before it enters the inferior dental canal, and in the anterior
portion of the canal it divides into the *incisive nerve* and the
mental nerve.

The **mylo-hyoid nerve,** accompanied by the mylo-hyoid
artery, lies in the mylo-hyoid groove.   It is distributed to the
mylo-hyoid muscle and to the anterior belly of the digastric
muscle.   The **incisive nerve** is distributed to the incisor teeth.

The **mental nerve,** accompanied by the mental artery,
passes through the mental foramen and is distributed to the
skin of the lower lip and of the chin, and to the mucous
membrane lining the lower lip.

As the inferior dental nerve lies in the inferior dental canal
it sends branches to the molar, bicuspid, and canine teeth.
(Morris, p. 783; Gray, p. 805.)

THE OTIC GANGLION.

The **otic ganglion** is situated in the course of the inferior maxillary division of the trifacial nerve, just outside the
foramen ovale.

The **motor root** of the otic ganglion comes from the nerve
to the internal pterygoid muscle.   The **sensory root** comes from

the facial nerve, as the small superficial petrosal nerve. The **sympathetic root** comes from the sympathetic plexus around the middle meningeal artery, as the small deep petrosal nerve.

The branches of the otic ganglion are distributed to the tensor tympani and the tensor palati muscles. It also sends communicating branches to the auriculo-temporal nerve and to the chorda tympani nerve. (Morris p. 787; Gray p. 807.)

THE SUBMAXILLARY GANGLION.

The **submaxillary ganglion** is situated, in the course of the inferior maxillary division of the trifacial nerve, on the superior surface of the submaxillary gland.

The **motor root** of the submaxillary ganglion comes from the chorda tympani nerve. The **sensory root** comes from the lingual nerve. The **sympathetic root** comes from the sympathetic plexus around the facial artery.

The branches of the submaxillary ganglion are distributed to the submaxillary gland and to Wharton's duct. (Morris p. 787; Gray p. 808.)

THE ABDUCENS NERVE.

The **superficial origin** of the abducens nerve is from the groove between the pons Varolii and the pyramid of the medulla. The **deep origin** of the abducens nerve is from a nucleus in the floor of the fourth ventricle. This nucleus is situated in the knee of the fibres of the facial nerve. The deep origin of the abduceus nerve is connected with the deep origin of the oculo-motor nerve by fibres which pass through the posterior longitudinal bundle (see p. 26). The abducens nerve is also connected with the sympathetic system and with the ophthalmic division of the trifacial nerve. The oculo-motor nerve and the trochlear nerve are also connected with the sympathetic system and with the ophthalmic division of the trifacial nerve.

The abducens nerve leaves the cranial cavity by passing through the sphenoidal fissure. It is distributed to the external rectus muscle of the eyeball. On account of the connection of the deep origin of this nerve with the deep origin of the oculo-motor, the external rectus muscle of one side and the internal rectus muscle of the opposite side act together. (Morris, p. 787; Gray, p. 810.)

THE FACIAL NERVE.

The **superficial origin** of the facial nerve is from the groove between the pons Varolii and the restiform body.

The **deep origin** of the facial nerve is from a nucleus deeply placed in the floor of the fourth ventricle. The fibres from this nucleus first pass toward the floor of the fourth ventricle, then bend upon themselves, forming the knee of the facial nerve, and lie around the deep origin of the abducens nerve. The fibres then pass through the medulla to emerge at the superficial origin of the nerve. The knee of the facial nerve forms a prominence, which is known as the eminentia teres, in the anterior half of the floor of the fourth ventricle.

The facial nerve leaves the cranial cavity by passing through the internal auditory meatus in company with the auditory nerve, the auditory artery, and the pars intermedia. At the bottom of the internal auditory meatus, the facial nerve enters the aqueductus Fallopii. It leaves the aqueductus Fallopii by passing through the stylo-mastoid foramen. It then passes through the parotid gland, at the anterior border of which it breaks up into its terminal branches.

The **pars intermedia** comes from the nucleus of origin of the glosso-pharyngeal nerve. It joins the facial nerve in the beginning of the aqueductus Fallopii.

The branches of the facial nerve are; (*a*) in the aqueductus Fallopii, (1) the *great superficial petrosal*, (2) the *small superficial petrosal*, (3) the *external petrosal*, (4) the *chorda tympani*, (5) the *nerve to the stapedius muscle*, and (6) the *communicating branch to the pneumogastric*; (*b*) just after leaving the stylo-mastoid foramen, (7) the *posterior auricular*, (8) the *nerve to the posterior belly of the digastric muscle*, and (9) the *nerve to the stylo-hyoid muscle;* (*c*) terminal branches, (10) *temporo-facial; temporal, malar,* and *infraorbital,* (11) *cervico-facial; buccal, supra-maxillary, and inframaxillary.* The terminal branches of the facial nerve form a radiating mass of nerves which is known as the **pes anserinus.**

The **great superficial petrosal nerve** is a branch of the facial in the aqueductus Fallopii. It leaves the aqueductus Fallopii and enters the cranial cavity by passing through the hiatus Fallopii. It passes across the petrous portion of the temporal bone and enters the cartilage which fills in the middle lacerated foramen. In this position it joins with the great deep petrosal nerve, to form the Vidian nerve, and goes to Meckel's ganglion as its motor root.

The **small superficial petrosal nerve** is a branch of the facial in the aqueductus Fallopii. It passes through a canal external to the hiatus Fallopii and enters the cranial cavity. It then passes through the foramen ovale and enters the otic ganglion as its sensory root. In its course through the canal in the petrous portion of the temporal bone it receives a large filament from the tympanic branch of the glosso-pharyngeal nerve.

The **external petrosal nerve** is a branch of the facial in the aqueductus Fallopii. It passes through a small canal in the petrous portion of the temporal bone and goes to the plexus of the sympathetic system around the middle meningeal artery.

[At this time it is convenient, for purposes of comparison, to give the courses of the deep petrosal nerves, although they have nothing to do with the facial nerve.

The **great deep petrosal nerve** is a branch of the carotid plexus of the sympathetic system. It enters the cartilage filling in the middle lacerated foramen and joins with the great superficial petrosal nerve to form the Vidian nerve. It then goes to Meckel's ganglion as its sympathetic root. (Morris, p. 845; Gray, p. 804.)

The **small deep petrosal nerve** is a branch of the sympathetic plexus around the middle meningeal artery. It forms the sympathetic root of the otic ganglion.

The **least deep petrosal nerve** is a branch of the tympanic branch of the glosso-pharyngeal nerve. It joins the carotid plexus of the sympathetic system. It is also called the **caroticotympanic** nerve.]

TABLE OF THE PETROSAL NERVES.

NERVE.	BRANCH OF.	DISTRIBUTED TO.
Great Superficial Petrosal.	Facial.	Meckel's ganglion, as its motor root.
Small Superficial Petrosal.	Facial; receives an important branch from Jacobson's nerve.	Otic ganglion, as its sensory root.
External Petrosal.	Facial.	Sympathetic plexus on middle meningeal artery.
Great Deep Petrosal.	Carotid plexus of sympathetic system.	Meckel's ganglion, as its sympathetic root.
Small Deep Petrosal.	Sympathetic plexus around middle meningeal artery.	Otic ganglion, as its sympathetic root.
Least Deep Petrosal.	Tympanic branch of glosso-pharyngeal.	Carotid plexus of sympathetic system.

The **chorda tympani nerve** is a branch of the facial in the aqueductus Fallopii. It leaves the aqueductus Fallopii and enters the tympanum by passing through the iter chordæ posterius. It passes through the tympanum, between the malleus and the incus, and enters the zygomatic fossa by passing through the iter chordæ anterius or the canal of Huguier. In the zygomatic fossa it joins with the lingual nerve and is distributed to the anterior portion of the tongue. The chorda tympani nerve is composed, principally, of the fibres of the pars intermedia and is probably concerned in taste.

The **nerve to the stapedius muscle** enters the tympanum by passing through a small canal which opens on the pyramid. It is distributed to the muscle the name of which it bears.

The **communicating branch to the pneumogastric nerve** joins the auricular branch of the pneumogastric as it passes through the auricular canal in the temporal bone.

The **posterior auricular nerve** is a branch of the facial, just outside the stylo-mastoid foramen. It lies between the external auditory meatus and the mastoid process of the temporal bone. It supplies the retrahens aurem, the attollens aurem, and the occipitalis muscles.

The **nerve to the posterior belly of the digastric muscle** and the **nerve to the stylo-hyoid muscle** supply the muscles indicated.

The **temporal branch** supplies the muscles of expression in the temporal region.

The **malar branch** supplies the muscles of expression around the orbit. The last two branches probably contain associated fibres from the oculo-motor nerves in order to make possible the coördinated movements of the lids with the eyeball.

The **infraorbital nerve** supplies the muscles of expression connected with the nose and with the upper lip. It forms the **infraorbital plexus** by anastomosing with the infraorbital portion of the superior maxillary division of the trifacial nerve.

The **buccal nerve** supplies the buccinator and orbicularis oris muscles.

The **supramaxillary nerve** supplies the muscles of expression connected with the lower lip.

The **inframaxillary nerve** supplies the platysma myoides muscle.

The branches of the facial nerve which supply the muscles of expression connected with the lips are probably associated with fibres from the hypoglossal nerve; so that coördinated movements between the tongue and the lips are possible. (Morris, p. 788; Gray, p. 811.)

THE AUDITORY NERVE.

The **superficial origin** of the auditory nerve is from the groove between the pons Varolii and the restiform body. The auditory nerve is external to the facial nerve in this situation.

The **deep origin** of the auditory nerve is from three nuclei, situated in the floor of the fourth ventricle. These nuclei are the accessory auditory nucleus or auditory ganglion, an external nucleus or Deiter's nucleus, and an internal nucleus or chief nucleus.

From these three nuclei the fibres which from the auditory nerve pass as a **lateral root** and as a **mesial root,** one on either side of the restiform body. The fibres of the lateral root come from the accessory auditory nucleus. The fibres of the mesial root come from the chief nucleus and from Deiter's nucleus.

Fibres pass from the accessory auditory nucleus to the chief nucleus and to Deiter's nucleus of the same side and of the opposite side. The crossed fibres are found in the striæ acusticæ (see page 31).

The **cortical area of hearing** is situated in the superior temporal convolution. The accessory auditory nucleus is connected to the superior temporal convolution in the following manner: fibres pass from the accessory nucleus or auditory ganglion, to the superior olive of the same side and of the opposite side; from the superior olive to the inferior corpus quadrigeminum, through the inferior fillet; from the inferior corpus quadrigeminum to the internal geniculate body, by the inferior brachium; and from the internal geniculate body to the superior temporal convolution.

The auditory nerve leaves the cranial cavity by passing through the internal auditory meatus, in company with the auditory artery, the facial nerve, and the pars intermedia.

At the bottom of the internal auditory meatus the auditory nerve divides into the *cochlear nerve* and the *vestibular nerve*.

The **cochlear nerve** is the continuation of the lateral root, before mentioned, and is the true nerve of hearing. It is distributed to the basement membrane which supports the organ of Corti (see Anatomy of Ear).

The **vestibular nerve** is distributed to the utricle and to the ampullæ of the semicircular canals. This branch is continuous with the mesial root and is probably concerned in the maintenance of equilibrium. (Morris p. 792; Gray p. 815.)

THE GLOSSO-PHARYNGEAL NERVE.

The **superficial origin** of the glosso-pharyngeal nerve is from the groove between the inferior olive and the restiform body. The glosso-pharyngeal is the most superior of the three nerves having their superficial origins from this groove.

The **deep origin** of the glosso-pharyngeal nerve is from the anterior part of the accessorio-vago-glosso-pharyngeal nucleus, the nucleus ambiguus, and the funiculus solitarius, in the floor of the fourth ventricle.

The **accessorio-vago-glosso-pharyngeal nucleus** is a sensory nucleus. The **nucleus ambiguus** is a motor nucleus and is, morphologically, a representative of the anterior horn of the gray matter of the spinal cord. The **funiculus solitarius** is a bundle of sensory fibres which begins in the spinal cord as low down as the decussation of the pyramids. The fibres have their origin from cells in the posterior horn of gray matter.

The glosso-pharyngeal nerve contains crossed motor fibres as well as fibres from the motor nucleus on the same side.

The **glosso-pharyngeal nerve** leaves the skull by passing through the middle compartment of the jugular foramen. It passes down the neck, lying between the internal jugular vein and the internal carotid artery, and then bends upon itself and runs forward, between the internal carotid and the external carotid arteries, and beneath the stylo-pharyngeus muscle. It lies beneath the hyo-glossus muscle and is distributed to the tongue and to the pharynx.

In its course the glosso-pharyngeal nerve presents a ganglion which is so constricted that two names are usually given to its different parts. The superior portion of this ganglion, situated in the jugular foramen, is called the **jugular ganglion;** the inferior portion is known as the **petrous ganglion.**

The branches of the glosso-pharyngeal nerve are: (1) the *meningeal,* (2) the *tympanic,* (3) the *tonsillar,* (4) the *muscular,* (5) the *pharyngeal,* and (6) the *lingual.*

The **meningeal branch** supplies the dura mater.

The **tympanic branch or Jacobson's nerve,** passes through the canalis tympanicus, in the petrous portion of the temporal bone, and enters the tympanum. In the tympanum it forms the tympanic plexus on the promontory. From the tympanic plexus an important branch is given off which joins the small superficial petrosal nerve as it passes through the petrous portion of the temporal bone.

The **muscular branch** supplies the stylo-pharyngeus muscle.

The **tonsillar branch** supplies the tonsil.

The **pharyngeal branches** go to the pharynx and, by uniting with branches from the pneumogastric and the sympathetic nerves, forms the **pharyngeal plexus.** The pharyngeal plexus rests on the middle constrictor muscle of the pharynx.

The **lingual branches** are distributed to the tongue, in the region of the circumvallate papillæ, and to the anterior surface of the epiglottis.

The glosso-pharyngeal nerve **communicates** with the sympathetic nerve, with the facial nerve, and with the carotid plexus of the sympathetic. The communicating branch from the glosso-pharyngeal nerve to the cartoid plexus of the sympathetic system is given off by the tympanic branch of the glosso-pharyngeal, and is known as the **least deep petrosal nerve.** (Morris, p. 794; Gray, p. 816.)

THE PNEUMOGASTRIC NERVE.

The **superficial** origin of the pneumogastric nerve is from the groove between the inferior olive and the restiform body. The pneumogastric is the middle of the three nerves having their superficial origins from this groove.

The **deep origin** of the pneumogastric nerve is from the middle portion of the accessorio-vago-glosso-pharyngeal nucleus, the nucleus ambiguus, and the funiculus solitarius, in the floor of the fourth ventricle.

The **pneumogastric nerve** leaves the skull by passing through the middle compartment of the jugular foramen; it then passes down the neck in the sheath with the internal carotid artery and the internal jugular vein; and with the common carotid artery and the internal jugular vein.   In the carotid sheath the pneumogastric nerve lies between and behind the vessels. A line drawn from the sterno-clavicular articulation to a point midway between the angle of the jaw and the mastoid process of the temporal bone would indicate the course of the nerve.

The **right pneumogastric nerve** then passes over the first portion of the subclavian artery and enters the thorax by passing behind the first rib.   It passes across the superior mediastinum, lying between the innominate artery and the right innominate vein, and runs backward to reach the posterior surface of the right bronchus.   It then goes to the esophagus and, after helping to form the esophageal plexus, passes behind that structure, through the esophageal opening in the diaphragm, to end in the solar plexus.

The **left pneumogastric** nerve passes between the left common carotid and the left subclavian arteries, behind the first rib, and across the superior mediastinum.   It passes in front of the arch of the aorta and then runs backward to the posterior surface of the left bronchus.   It then goes to the esophagus and, after helping to form the esophageal plexus, passes in front of that structure, through the esophageal opening in the diaphragm, to end on the anterior surface of the stomach.

As the pneumogastric nerve passes through the jugular foramen it presents an enlargement known as the **ganglion of the root.**

The **ganglion of the trunk** is a ganglionic enlargement in the course of the pneumogastric nerve, just after it has emerged from the jugular foramen. The hypoglossal nerve winds around this ganglion in a spiral manner, from within outward. The accessory portion of the spinal accessory nerve joins this ganglion.

The branches of the pneumogastric nerve are: (1) the *meningeal*, (2) the *auricular*, (3) the *pharyngeal*, (4) the *superior laryngeal*, (5) the *superior cervical cardiac*, (6) the *inferior cervical cardiac*, (7) the *inferior or recurrent laryngeal*, (8) the *thoracic cardiac*, (9) the *anterior pulmonary*, (10) the *posterior pulmonary*, (11) the *esophageal*, (12) the *gastric*, (13) the *splenic*, and (14) the *hepatic*.

The **meningeal branch** supplies the dura mater.

The **auricular branch or Arnold's nerve**, passes through the canalis auricularis, in the petrous portion of the temporal bone, and is distributed to the external auditory meatus and to the auricle.

The **pharyngeal branch** passes forward, between the external carotid and the internal carotid arteries, and, on the middle constrictor muscle of the pharynx, helps to form the pharyngeal plexus.

The **superior laryngeal nerve** passes inward and enters the pharynx, in company with the superior laryngeal artery, by piercing the thyro-hyoid membrane. It is distributed to the mucous membrane of the larynx. Just before the superior laryngeal nerve pierces the thyro-hyoid membrane it gives off the **external laryngeal nerve,** which supplies the crico-thyroid muscle.

The **superior cervical cardiac nerve** passes downward and goes to the deep cardiac plexus.

The **inferior cervical cardiac nerve of the right side** goes to the deep cardiac plexus. The corresponding nerve of the **left** side goes to the superficial cardiac plexus.

The **inferior or recurrent laryngeal nerve** differs in its point of origin on the two sides of the body. The **right nerve** is given off as the pneumogastric nerve crosses the right subclavian artery. It winds around the subclavian artery and enters the neck. The **left nerve** is given off as the pneumogastric nerve crosses the arch of the aorta. It winds around the arch of the aorta and passes upward through the superior mediastinum, behind the left common carotid artery, and enters the neck. In the neck, both nerves lie in the groove between the trachea and the esophagus. They enter the larynx by passing beneath the inferior constrictor muscle of the pharynx and behind the crico-thyroid articulation. The inferior laryngeal nerve supplies all the muscles of the larynx, except the crico-thyroid. It also gives off a branch which is distributed to the mucous membrane of the larynx.

The **thoracic cardiac branches** go to the deep cardiac plexus.

The **anterior pulmonary branches** communicate with branches from the sympathetic nerve and form the posterior pulmonary plexus. This plexus lies on the anterior surface of the bronchus.

The **posterior pulmonary branches** ramify over the posterior surface of the bronchus and form the posterior pulmonary plexus by communicating with sympathetic fibres.

The **esophageal branches** form the esophageal plexus.

The **gastric branches** come from the left pneumogastric nerve and accompany the gastric artery in its course along the lesser curvature of the stomach. Gastric branches come from the right pneumogastric nerve, also, and are distributed to the posterior wall of the stomach.

The **splenic branches** are given off from the right pneumogastric nerve and accompany the splenic artery.

The **hepatic branches** are given off from the left pneumogastric nerve and accompany the hepatic artery.

The pneumogastric nerve **communicates** with the facial, the glosso-pharyngeal, the spinal accessory, the hypoglossal, the first and second cervical, and the sympathetic nerves. (Morris, p. 796; Gray, p. 819.)

THE SPINAL ACCESSORY NERVE.

The spinal accessory nerve is composed of two distinct parts; a *spinal portion* and an *accessory portion*.

The **superficial origin of the accessory portion** of the spinal accessory nerve is from the groove between the inferior olive and the restiform body. The accessory portion of the spinal accessory nerve is the most inferior of the three nerves having their superficial origins from this groove.

The **deep origin of the accessory portion** of the spinal accessory nerve is from the posterior part of the accessorio-vago-glosso-pharyngeal nucleus, the nucleus ambiguus, and the funiculus solitarius, in the floor of the fourth ventricle.

The **superficial origin of the spinal portion** of the spinal accessory nerve is from the lateral aspect of the spinal cord, between the ligamentum denticulatum and the posterior root of the spinal nerve, as low down as the seventh cervical nerve.

The **deep origin of the spinal portion** of the spinal accessory nerve is from a group of cells in the anterior horn of gray matter of the spinal cord.

The spinal portion of the spinal accessory nerve enters the cranial cavity by passing through the foramen magnum, in the occipital bone. It then joins with the accessory portion and the two portions leave the cranial cavity as one nerve, by passing through the middle compartment of the jugular foramen. As soon as the nerve leaves the jugular foramen, the accessory portion joins the ganglion of the trunk of the pneumogastric nerve and goes, principally, to the pharyngeal plexus, through the pharyngeal branch of the pneumogastric nerve. The spinal portion then pierces the sterno-mastoid muscle, supplies it, and passes across the occipital triangle, beneath the anterior border of the trapezius muscle to supply that muscle.

The **sterno=mastoid plexus** is formed, beneath the sterno-mastoid muscle, by fibres from the second cervical nerve and fibres from the spinal accessory nerve.

The **subtrapezial plexus** lies beneath the trapezius muscle. It is formed by fibres from the spinal accessory nerve and fibres from the third and fourth cervical nerves. (Morris, p. 800; Gray, p. 823.)

THE HYPOGLOSSAL NERVE.

The **superficial origin** of the hypoglossal nerve is from the groove between the inferior olive and the pyramid of the medulla.

The **deep origin** of the hypoglossal nerve is from the hypoglossal nucleus in the floor of the fourth ventricle. The two hypoglossal nuclei are connected by commissural fibres. The nucleus of origin of the hypoglossal nerve and the nuclei of origin of the facial and oculo-motor nerves are connected by fibres which pass through the posterior longitudinal bundle.

The **hypoglossal nerve** leaves the cranial cavity by passing through the anterior condyloid foramen. After emerging from the anterior condyloid foramen, the hypoglossal nerve lies; first to the inner side, then behind, then to the outer side, and finally, in front of the ganglion of the trunk of the pneumogastric nerve. It then hooks around the occipital artery and runs forward, lying superficial to the external carotid artery. It then passes across the middle constrictor muscle of the pharynx, lies in front of the hyo-glossus muscle, and breaks up into its terminal branches under cover of the mylo-hyoid muscle. In its course the hypoglossal nerve lies above the lingual artery and the greater cornu of the hyoid bone.

The branches of the hypoglossal nerve are: (1) the *meningeal*, (2) the *descendens hypoglossi*, (3) the *nerve to the stylo-hyoid muscle*, (4) the *nerve to the hyo-glossus muscle*, and (5) the *nerve to the genio-hyo-glossus muscle*.

The **meningeal branch** supplies the dura mater of the brain.

The **descendens hypoglossi** nerve is given off from the hypoglossal nerve, just as the latter nerve winds around the occipital artery. It passes downward, lying in front of the sheath of the carotid vessels and joins with the communicans hypoglossi nerves, from the second and third cervical nerves, to form the **ansa hypoglossi.** From the descendens hypoglossi nerve itself or from the ansa hypoglossi, branches are given off which supply the omo-hyoid, sterno-hyoid, sterno-thyroid, thyro-hyoid, and genio-hyoid muscles.

The hypoglossal nerve is joined by branches from the first and second cervical nerves as it lies in front of the gang-lion of the trunk of the pneumogastric nerve. The spinal fibres thus associated with the hypoglossal nerve leave it as the descendens hypoglossi branch.

The hypoglossal nerve receives **communicating branches** from the lingual branch of the trifacial nerve and from the pneumogastric nerve. (Morris, p. 801 ; Gray, p. 823.)

THE SUPERFICIAL AND DEEP ORIGINS OF THE CRANIAL NERVES.

NERVE.	SUPERFICIAL ORIGIN.	DEEP ORIGIN.
Olfactory.	External root from uncinate gyrus, and amygdaloid nucleus ; internal root from anterior extremity of gyrus fornicatus (origin of olfactory tract).	Cortex of olfactory bulb.
Optic.	Outer side of crus cerebri.	Optic thalamus, external geniculate body, and anterior corpus quadrigeminum.
Oculo-motor.	Inner side of crus cerebri.	Anterior part of floor of aqueduct of Sylvius.
Trochlear.	Valve of Vieussens (anterior medullary velum).	Posterior part of floor of aqueduct of Sylvius.
Trifacial.	By a motor root and a sensory root from the side of the pons Varolii.	A motor nucleus in the floor of the fourth ventricle, assisted by a descending root from the floor of the aqueduct of Sylvius, and a sensory nucleus in the floor of the fourth ventricle, assisted by an ascending root from the medulla.
Abducens.	The groove between the pons and the pyramid of the medulla.	A nucleus in the floor of the fourth ventricle.
Facial.	The groove between the pons and the restiform body.	A nucleus in the floor of the fourth ventricle.

THE SUPERFICIAL AND DEEP ORIGINS OF THE CRANIAL NERVES.

NERVE.	SUPERFICIAL ORIGIN.	DEEP ORIGIN.
Auditory.	The groove between the pons and the restiform body.	The auditory ganglion or accessory auditory nucleus (cochlear branch), and Deiter's nucleus and the chief nucleus (vestibular branch). The three nuclei are situated in the floor of the fourth ventricle.
Glosso-pharyngeal.	The groove between the inferior olive and the restiform body.	The anterior portion of the accessorio-vago-glosso-pharyngeal nucleus, the nucleus ambiguus, and the funiculus solitarius. In the floor of the fourth ventricle.
Pneumogastric	The groove between the inferior olive and the restiform body.	The middle portion of the accessorio-vago-glosso-pharyngeal nucleus, the nucleus ambiguus, and the funiculus solitarius. In the floor of the fourth ventricle.
Spinal accessory.	The groove between the inferior olive and the restiform body (accessory portion). The side of the spinal cord, between the posterior roots of the spinal nerves and the ligamentum denticulatum (spinal portion).	The posterior portion of the accessorio-vago-glosso-pharyngeal nucleus, the nucleus ambiguus, and the funiculus solitarius (accessory portion). Cells in the anterior horn of gray matter of the spinal cord as low down as the seventh cervical nerve (spinal portion).
Hypoglossal.	The groove between the inferior olive and the pyramid of the medulla.	A nucleus in the floor of the fourth ventricle.

CHAPTER IV.

THE SYMPATHETIC NERVES.

The **sympathetic nerves** are distributed to the involuntary muscle in the body. They are not distinct nerves; but are associated closely with the spinal nerves.

The sympathetic system is composed of *the gangliated cords* and of *the prevertebral plexuses*.

THE GANGLIATED CORDS.

The **gangliated cords** are situated on either side of the vertebral column, and originally presented a pair of ganglia for each pair of spinal nerves. In the cervical, lumbar and sacral regions, certain of the ganglia become fused, so that, in the adult, we are able to isolate only about twenty-three pairs.

These ganglia are connected to the spinal nerves by the **rami communicantes.** There are usually two rami communicantes to each ganglion. One of these branches is **white,** composed of medullated nerve fibres, which come from the spinal nerves and which go to the sympathetic nerves. The other branch is **gray,** composed of non-medullated nerve fibres, which come from the sympathetic ganglia and which go to the spinal nerves to be distributed principally to the blood vessels and glands in the course of these nerves.

The fibres which come from the spinal nerves to the sympathetic may have one of the following courses: first, a fibre may pass to the nearest ganglion and form a terminal arborization around one of the nerve cells in that ganglion; second, a fibre may pass through the nearest ganglion to be interrupted in relation with a cell in a ganglion farther along the gangliated cord; third, a fibre may pass through a series of ganglia and on to one of the prevertebral plexuses before it is interrupted: fourth, a fibre may pass through a series of

79

ganglia, and through one of the prevertebral plexuses to be interrupted finally in one of the ganglia in a viscus.

The cells in the ganglia of the gangliated cord may send fibres directly to the viscera or fibres may be interrupted in relation with cells in the prevertebral plexuses and new fibres pass thence to the viscera.

In the **cervical region** the gangliated cords present three pairs of ganglia; the *superior*, the *middle*, and the *inferior cervical ganglia*.

The **superior cervical ganglion** rests on the transverse processes of the second and third cervical vertebræ. It is formed by the fusion of the first four ganglia.

The branches of the superior cervical ganglion may be divided into an *ascending group*, a *descending group*, an *internal group*, an *external group*, and an *anterior group*.

The **ascending branches** of the superior cervical ganglion pass upward along the internal carotid artery, and are so distributed that they form two plexuses; the *cavernous plexus* and the *carotid plexus*.

The **cavernous plexus** is situated on the inner side of the internal carotid artery, after that artery has entered the cranial cavity. Branches of this plexus are distributed to the branches of the internal carotid artery; others pass as communicating branches to the oculo-motor nerve, the trochlear nerve, and the ophthalmic division of the trifacial nerve; and a third set form the sympathetic root of the ophthalmic ganglion.

The **carotid plexus** is found on the outer side of the internal carotid artery. It gives off the following branches: (1) the *great deep petrosal*, (2) the *least deep petrosal*, (3) *branches to the Gasserian ganglion*, and (4) *communicating branches to the abducens nerve*.

The **great deep petrosal nerve** enters the cartilage which fills in the middle lacerated foramen and joins with the great superficial petrosal nerve to form the Vidian nerve. The fibres of this nerve then proceed to Meckel's ganglion as its sympathetic root.

The **least deep petrosal nerve** joins the tympanic plexus. The **descending branch** of the superior cervical ganglion is the **superior cardiac nerve.** The superior cardiac nerve of the right side goes to the deep cardiac plexus. The left superior cardiac nerve goes to the superficial cardiac plexus.

The **internal branch** of the superior cervical ganglion unites with the pharyngeal branches of the pneumogastric and of the glosso-pharyngeal nerves to form the **pharyngeal plexus.** This plexus lies on the middle constrictor muscle of the pharynx.

The **external branches** pass to the glosso-pharyngeal, the pneumogastric, the hypoglossal, and the spinal nerves, as communicating branches.

The **anterior branches** are known as the **nervi molles.** They are distributed to the external carotid artery, forming plexuses on the walls of each of its branches.

The **middle cervical ganglion** lies opposite to the transverse process of the sixth cervical vertebra, in front of the inferior thyroid artery. It is formed by the fusion of the fifth and sixth ganglia. It gives off the *middle cardiac nerve* and *branches to the thyroid body.*

The **middle cardiac nerves** go to the deep cardiac plexus.

The **inferior cervical ganglion** is formed by the fusion of the seventh and eighth ganglia. It lies between the neck of the first rib and the transverse process of the seventh cervical vertebra. It gives off the *inferior cardiac nerve* and branches which ramify on the vertebral artery to form the *vertebral plexus.*

The **vertebral** plexus is continued into the cranial cavity and forms the basilar plexus on the basila arrtery.

The **inferior cardiac nerves** go to the deep cardiac plexus.

In the **thoracic region** there are twelve pairs of ganglia. The upper ten ganglia lie on the heads of the first ten ribs; the lower two ganglia lie on the sides of the bodies of the last two thoracic vertebræ.

The **first five ganglia** give off branches which pass to the *esophagus,* the *thoracic aorta,* the *mediastinum,* and the *lungs.* The **ganglia from the fifth to the twelfth,** inclusive, give off branches which form the *great splanchnic nerve*, the *lesser splanchnic nerve*, and the *least splanchnic nerve.*

The **great splanchnic nerve** is formed by branches from the fifth, sixth, seventh, eighth, and ninth thoracic ganglia. It passes behind the crus of the diaphragm and ends in the semilunar ganglion.

The **lesser splanchnic nerve** is formed by branches from the tenth and eleventh thoracic ganglia. It passes behind the crus of the diaphragm and goes to the solar plexus.

The **least splanchnic nerve** is formed by a branch from the twelfth thoracic ganglion. It pierces the diaphragm and goes to the renal plexus.

There are four pairs of ganglia in the **lumbar portion** of the sympathetic nerve. These ganglia lie in front of the bodies of the lumbar vertebræ. They give branches to the aortic plexus.

There are four pairs of ganglia in the **sacral portion** of the sympathetic nerve. These ganglia lie on the anterior surface of the sacrum, internal to the anterior sacral foramina. They give branches to the pelvic plexus.

The right and the left gangliated cords of the sympathetic system terminate in the **ganglion impar,** which is situated in front of the first segment of the coccyx.

THE PREVERTEBRAL PLEXUSES.

The **deep cardiac plexus** lies between the arch of the aorta and the bifurcation of the trachea. It is formed by the right superior cardiac branch, the right and left middle cardiac branches, and the right and left inferior cardiac branches of the cervical sympathetic nerve, and by the right and left superior cervical cardiac branches, the right inferior cervical cardiac branch, and the right and left thoracic cardiac branches of the pneumogastric nerves. The deep cardiac plexus sends branches to the anterior pulmonary plexus and to the plexuses surrounding the coronary arteries.

The **superficial cardiac plexus** lies in the concavity of the arch of the aorta, just above the pulmonary artery. It is formed by the left superior cardiac branch of the cervical sympathetic nerve and the inferior cervical cardiac branch of the left pneumogastric nerve. The superficial cardiac plexus sends branches to the plexus surrounding the right coronary artery.

The **solar plexus** is situated behind the stomach and in front of the celiac axis. It is formed by the right and left semilunar ganglia, the right and left great splanchnic nerves, the right and left lesser splanchnic nerves, and the right pneumogastric nerve. The solar plexus gives off branches, which follow the various arteries arising from the abdominal aorta to form plexuses on their walls. These plexuses take their names from the arteries with which they are in relation.

The **phrenic plexus** is formed by branches from the solar plexus. It passes with the phrenic arteries, supplies the diaphragm, and sends filaments to the inferior vena cava. Communicating branches come to it from the phrenic nerve.

The **celiac plexus** surrounds the celiac axis. It divides into the *hepatic plexus,* the *splenic plexus,* and the *gastric plexus.*

The **hepatic plexus** is formed by branches from the solar plexus and from the left pneumogastric nerve.

The **splenic plexus** is formed by branches from the solar plexus and from the right pneumogastric nerve.

The **gastric plexus** is formed by branches from the solar plexus.

The **suprarenal plexus** is formed by branches from the phrenic plexus, the solar plexus, and the renal plexus. It is distributed to the suprarenal body.

The **renal plexus** is formed by the least splanchnic nerve and by branches from the solar plexus and from the aortic plexus. It is distributed to the kidney.

The **spermatic plexus** is formed by branches from the renal plexus and from the aortic plexus. The corresponding plexus in the female is known as the **ovarian plexus.** The spermatic plexus supplies the testicle. The ovarian plexus supplies the ovary.

The **superior mesenteric plexus** is formed by branches from the solar plexus.

The **aortic plexus** is formed by branches from the solar plexus and by branches from the lumbar ganglia.

The **inferior mesenteric plexus** is formed by branches from the aortic plexus.

The **hypogastric plexus** lies in front of the body of the fifth lumbar vertebra, in the bifurcation of the ab-dominal aorta. It is formed by the continuation of the aortic plexus and by branches from the lower lumbar ganglia. It divides into two branches, the *right and left pelvic plexuses.*

The **pelvic plexuses** are situated on either side of the rectum and are formed by the two branches of the hypogastric plexus and by branches from the sacral ganglia. The pelvic plexuses give off branches which form the *vesical plexus,* the *hemorrhoidal plexus,* the *prostatic plexus,* the *uterine plexus,* and the *vaginal plexus.*

The **vesical plexus,** in the male, sends branches to the vas deferens and to the seminal vesicles.

The **prostatic plexus,** seen only in the male, sends fila-ments to the corpora cavernosa and to the corpus spongiosum.

The **uterine plexus** receives filaments from the ovarian plexus.

The **vaginal plexus** receives branches from the sacral nerves. (Morris, p. 843; Gray, p. 867.)

CHAPTER V.

THE EAR.

The auditory apparatus is composed of (1) the *external ear*, (2) the *middle ear or tympanum*, and (3) the *internal ear*.

The **external ear** is composed of the *auricle or pinna* and the *external auditory canal*.

The ground substance of the **auricle or pinna** is composed of yellow elastic cartilage, which is covered over by skin. The pinna presents the following points for examination: a prominent external ridge or **helix** which is separated by a groove, the **fossa of the helix,** from a less prominent **antihelix.** The superior extremity of the antihelix bifurcates into two limbs, which enclose between them a triangular depression known as the **fossa of the antihelix.** The **tragus** is a prominent tubercle situated just below the inner extremity of the helix; the **antitragus** is a smaller tubercle situated opposite to the tragus at the lower extremity of the antihelix. The **lobule** is the most dependent portion of the pinna; it contains no cartilage. The concave portion of the pinna which is situated immediately in front of the external auditory canal is known as the **concha.**

The pinna is connected to the zygoma by the **anterior ligament** and to the mastoid process by the **posterior ligament.**

The **intrinsic muscles** of the pinna are the *helicis major,* the *helicis minor,* the *tragicus,* the *antitragicus,* the *transversus auris* and the *obliquus auris.*

The **extrinsic muscles** of the pinna are the *attrahens aurem,* the *attollens aurem,* and the *retrahens aurem.*

The **vessels of the pinna** are branches of the superficial temporal artery, of the posterior auricular artery, and of the occipital artery.

The **nerves of the pinna** are branches of the auriculo-temporal nerve, of the auricularis magnus nerve, and of the posterior auricular branch of the facial nerve. The auricular branch of the pneumogastric nerve is also distributed to the pinna.

The **external auditory canal** is composed partly of cartilage and partly of bone. The bony portion of the external auditory canal belongs to the temporal bone. The **fissures of Santorini** are small, linear defects in the cartilaginous portion of the external auditory canal. The external auditory canal is lined by skin, which contains modified sebaceous and sudoriferous glands. These glands secrete the cerumen.

The external auditory canal is directed at first forward and upward, then horizontally, and finally forward and downward; so that, in order to bring the bottom of the canal into view, the pinna must be pulled upward and backward.

The bottom of the external auditory canal is separated from the tympanum by the **tympanic membrane.** The tympanic membrane is an ellipsoid structure composed of three layers of tissue; on its outer aspect, it is covered by skin; on its inner surface, it is lined by the mucous membrane of the tympanum; and between these two layers there is a layér of fibrous tissue, into which the manubrium of the malleus is inserted. The tympanic membrane is inserted into the bony tympanic ring in such a manner that it forms an angle of about 55° with the external auditory canal. This ring is deficient in its superior portion and there is a robust band of tissue attached to either side of the gap. Between these two dense ligaments there is a loosely stretched portion of the tympanic membrane spoken of as the **membrana flaccida** or **Shrapnell's membrane.** In examining the tympanic membrane, by the aid of the head mirror and an appropriate speculum, the following markings may be seen: *Shrapnell's membrane,* superiorly; a rounded prominence made by the *processus brevis of the malleus;* a long prominence, indicated by an area highly refractive to light, due to the underlying *manubrium of the malleus;* and the *umbo,* produced by the attachment of the handle of the malleus to the tympanic membrane. (Morris, p. 886; Gray, p. 912.)

The **middle ear or tympanum** is a space situated in the substance of the petrous portion of the temporal bone. It is bounded; externally, by the tympanic membrane; internally, by the petrous portion of the temporal bone; superiorly, by the tegmen tympani; inferiorly, by the plate of bone which forms the roof of the jugular fossa; and posteriorly, by the mastoid portion of the temporal bone. The roof and the floor approach each other anteriorly and posteriorly. Anteriorly the internal orifice of the **canalis musculo-tubaris** is to be seen, which transmits the Eustachian tube and the tendon of the tensor tympani muscle. The two structures are separated from each other by the **processus cochleariformis.** The **promontory,** which is formed by the first turn of the cochlea, may be seen on the inner wall of the tympanum. The **oval window** is seen just above the promontory; the **round window** is just below the promontory. The oval window opens into the vestibule, it is closed by the base of the stapes; the round window opens into the scala tympani of the cochlea, it is closed by the secondary tympanic membrane. At the posterior portion of the tympanum there is a prominence produced by the aqueductus Fallopii, which is known as the **pyramid.**

The cavity of the tympanum may be subdivided into the *atrium,* the *attic,* and the *antrum.* The **atrium** is that portion of the tympanic cavity below the superior margin of the external auditory canal. The **attic** is that portion of the tympanic cavity above the superior margin of the external auditory canal. The **antrum** is that portion of the tympanic cavity which opens into the mastoid cells. The **mastoid cells** are air cells in the mastoid portion of the temporal bone. They are separated from the lateral sinus by a thin plate of bone.

The tympanum contains the *three ear ossicles, two muscles,* and *two nerves.*

The **ear ossicles** are three in number, the *malleus,* the *incus,* and the *stapes.* The **malleus** articulates, by its head, with the incus. The **manubrium** or handle of the malleus is inserted into the tympanic membrane; the **processus gracilis** of the malleus is contained in the Glasserian fissure; the **processus brevis** impinges against the tympanic membrane.

The **incus** articulates, by its body, with the head of the malleus. The **long process** of the incus lies almost parallel with the manubrium of the malleus and articulates with the head of the stapes. The **short process** of the incus projects backward into the tympanum.

The head of the **stapes** articulates with the long process of the incus. The head is connected to the base of the stapes by the **arch of the stapes.** The **base** of the stapes is movably attached to the margin of the oval window.

The **tensor tympani muscle** arises from the cartilaginous portion of the Eustachian tube and from the adjacent portions of the sphenoid bone. Its tendon passes through the superior portion of the canalis musculo-tubaris, and is inserted into the manubrium of the malleus.

The **stapedius muscle** arises from the pyramid and is inserted into the neck of the stapes.

The **tympanic branch of the glosso-pharyngeal nerve** forms the tympanic plexus on the promontory of the tympanum.

The **chorda tympani nerve** passes across the tympanum, between the manubrium of the malleus and the incus.

The **Eustachian tube** begins on the anterior portion of the floor of the tympanum and passes downward, forward, and inward, through the lower compartment of the canalis musculo-tubaris, to open on the posterior wall of the pharynx. As the Eustachian tube passes through the canalis musculo-tubaris it is separated from the tendon of the tensor tympani muscle by a thin plate of bone, known as the processus cochleariformis. The Eustachian tube is partly bony and partly cartilaginous. Its mucous membrane is covered by ciliated columnar epithelium. (Morris, pp. 56 and 889; Gray, p. 916.)

The **internal ear or labyrinth** consists of a *bony portion* and a *membranous portion*.

The **bony labyrinth** is imbedded in the substance of the petrous portion of the temporal bone, so that its long axis lies parallel to the axis of that bone. It is divided into the *vestibule*, the *semicircular canals*, and the *cochlea*.

The three **semicircular canals** lie behind the vestibule. They are named the *superior*, the *posterior*, and the *external*

semicircular canals, and lie in three distinct planes. Each semicircular canal has a small extremity and a dilated extremity or
ampulla. The small ends of the superior and the posterior
semicircular canals unite and open into the vestibule by a
common aperture. The three ampullæ open into the vestibule
by distinct apertures. There are, therefore, five orifices leading
from the posterior part of the vestibule.

The **vestibule** is the middle portion of the bony labyrinth; the semicircular canals open from it posteriorly and
the cochlea opens from it anteriorly. The opening of the
oval window is to be seen on the outer wall of the
vestibule. The **fovea hemispherica** is a rounded depression on the inner wall of the vestibule; it is perforated by
numerous foramina for the passage of the filaments of the
cochlear branch of the auditory nerve. The **crista vestibuli**
is a bony ridge situated between the fovea hemispherica and
the fovea hemielliptica. The **fovea hemielliptica** is an ellipsoid
depression on the superior wall of the vestibule; it is perforated
by numerous small foramina for the passage of the branches
of the vestibular branch of the auditory nerve.

The inner wall of the vestibule is in relation with the
cribriform plate at the bottom of the internal auditory meatus.
The **falciform crest** passes across this plate of bone and
divides it into a superior and an inferior portion. The superior
portion of the cribriform plate is perforated by the numerous
small foramina which are seen in the fovea hemielliptica. The
inferior portion of the cribriform plate is perforated by two groups
of foramina; one of these groups is seen in the fovea hemispherica and the other transmits the nerves to the cochlea.

The **cochlea** lies in front of the vestibule and has an
axis which is nearly at right angles with the axis of the petrous
portion of the temporal bone. The base of the cochlea is
directed inward and the apex or **cupola** is directed outward.
The cochlea consists of two and one-half turns around a central
bony **modiolus.** The base of the cochlea is in relation with
the cribriform plate at the bottom of the internal auditory meatus.

The bony canal of the cochlea is incompletely divided by a

plate of bone, known as the **lamina spiralis,** which projects
into it from the modiolus. The modiolus and the lamina
spiralis are pierced by canals for the passage of the filaments
of the cochlear nerve.

The portion of the osseous canal above the lamina spiralis
is known as the **scala vestibuli;** the portion below the lamina
spiralis is known as the **scala tympani.** The scala vestibuli
opens into the vestibule; the scala tympani opens into the
tympanum, through the **round window.** The scala vestibuli
and the scala tympani communicate at the cupola by a channel
known as the **helicotrema.**

The **membranous labyrinth** is divided into the *vestibule,*
the *semicircular canals,* and the *cochlea.* The space between
the bony labyrinth and the membranous labyrinth is lined by
endothelial cells and is filled by a modified lymph, known as
the **perilymph.**

The **membranous vestibule** is subdivided into a posterior
portion or **utricle,** which is in relation with the fovea hemielliptica, and an anterior portion or **saccule,** which is in relation
with the fovea hemispherica. The utricle communicates with
the saccule by a Y-shaped duct, known as the **ductus endolymphaticus.** The common arm of this duct passes through
the aqueductus cochleæ, in the petrous portion of the temporal
bone, and ends, in the **saccus endolymphaticus** beneath the
dura mater.

The **membranous semicircular canals** lie within the bony
semicircular canals. Each canal presents a dilated end or **ampulla** and a smaller end. They open directly into the utricle.
The smaller ends of the superior and the posterior semicircular
canals open by a common orifice.

The semicircular canals are lined by epithelium, which, in
the ampullæ, is specialized to receive impulses concerned in the
maintenance of equilibrium. This area is termed the **macula
acustica** and the epithelium here is a variety of neuro-epithelium,
having long hair-like processes. Above the hair cells, in the
ampullæ of the semicircular canals, numerous crystals of car-
bonate of lime, known as **otoliths,** are to be found.

The **membranous cochlea** opens into the saccule by a short passage termed the **canalis reuniens.** The membranous cochlea holds such a relation to the bony cochlea that the partition partly formed by the lamina spiralis is completed. The membranous cochlea, then, lies between the scala tympani and the scala vestibuli and is known as the **scala media.**

The scala media is separated from the scala tympani by a membrane which passes from the tip of the lamina spiralis, straight across to the external wall of the bony cochlea; this membrane is known as the **basilar membrane.** The scala media is separated from the scala tympani by a membrane which passes from the superior surface of the lamina spiralis, obliquely upward and outward, to the external wall of the bony cochlea; this membrane is known as the **membrane of Reissner.**

The scala media is lined by epithelium which at one point is specialized to receive auditory impulses. The **organ of Corti** is the name given to that portion of the epithelium which receives the auditory impulses. It is situated on the basilar membrane and is composed of neuro-epithelial cells and of sustentacular cells. In a transverse section of the organ of Corti these elements are arranged as follows: two of the sustentacular elements are more robust than their fellows and lean toward each other, so that they come in contact at their upper extremities. These elements are known as the **inner** and **outer pillars of Corti;** they include between them a triangular space to which the name of the **tunnel of Corti** has been given. Resting against the pillars of Corti the neuro-epithelial elements may be seen. They are columnar cells which rest, below, on the basilar membrane and which present, on their free extremities, several hair-like, protoplasmic processes. The presence of these processes has caused these cells to be known as the **hair-cells.** These hair-cells are so arranged that, on transverse section, one cell is to be seen internal to the inner pillar of Corti, the **internal hair-cell;** and three cells are to be seen external to the external pillar of Corti, the **external hair-cells.** The free extremities of the hair-cells are held in apposition by the **reticular membrane,** which is formed by certain of the sustentacular cells, known as the **cells of Deiters.**

The cilia project through small apertures in the reticular membrane. Above the cilia we find a membrane which acts as a damper for after vibrations; it is termed the **tectorial membrane.**

The membranous labyrinth is filled with a modified lymph, known as the **endolymph,** the vibrations of which make impressions on the cilia of the hair-cells. (Piersol, p. 383; Morris, p. 895; Gray, p. 921.)

THE DEVELOPMENT OF THE EAR.

The **auricle** is developed in the tissues of the first and second visceral arches around the first visceral furrow.

The **external auditory canal** is developed from the first visceral furrow.

The **tympanum** and the **Eustachian tube** develop from the first pharyngeal pouch.

The **tympanic membrane** is formed from the bridge of tissue which separates the first visceral furrow from the first pharyngeal pouch.

The **malleus** and the **incus** are developed from the rod of cartilage contained in the first visceral arch (Meckel's cartilage).

The **stapes** develops from the rod of cartilage contained in the second visceral arch.

The **membranous labyrinth** is formed by the ingrowth of the ectoderm of the surface of the body. This ingrowth forms a canal which lies opposite the position of the medulla oblongata; it is called the **otic vesicle.** By processes of constriction and unequal growth the various parts of the membranous labyrinth, utricle, saccule, semicircular canals, and scala media of the cochlea, are formed. The otic vesicle grows into the underlying mesoderm which subsequently becomes hollowed out to form the space between the membranous labyrinth and the bony labyrinth. In the cochlea these spaces become; the **scala tympani,** below; and the **scala vestibuli,** above the developing membranous cochlea. The mesodermic tissue around these spaces soon takes on the characteristic appearance of cartilage and this cartilage subsequently ossifies. (Quain, p. 89; A. T. O., p. 132.)

CHAPTER VI.

THE EYE.

The eye and its appendages are contained in the orbit.

The **orbit** is a pyramidal cavity, the base of which is directed outward and the apex backward. The **outer wall of the orbit** is formed by the orbital plate of the great wing of the sphenoid bone and by the orbital process of the malar bone. The **inner wall of the orbit** is formed by the lachrymal bone, the os planum of the ethmoid bone, and part of the body of the sphenoid bone. The **roof of the orbit** is formed by the orbital plate of the frontal bone and by the lesser wing of the sphenoid bone. The **floor of the orbit** is formed by the orbital plate of the superior maxillary bone, the orbital process of the malar bone, and the orbital process of the palate bone.

The **base of the orbit** is surrounded by a prominent ridge of bone formed by the frontal bone, the nasal process of the superior maxillary bone, the body of the superior maxillary bone, and the malar bone.

At the apex of the orbit, the **optic foramen** opens for the transmission of the optic nerve and the ophthalmic artery. Just external to the optic foramen the opening of the **sphenoidal fissure** is to be seen. The sphenoidal fissure transmits the oculo-motor nerve, the trochlear nerve, the ophthalmic division of the trifacial nerve, the abducens nerve, filaments of the cavernous plexus of the sympathetic nerve, the recurrent branch of the lachrymal artery, the orbital branch of the middle meningeal artery, and the ophthalmic vein. Between the outer wall and the floor of the orbit the **spheno-maxillary** fissure is to be seen. It transmits the superior maxillary division of the trifacial nerve, the infraorbital artery, and the ascending branches of Meckel's ganglion. The spheno-maxillary fissure forms almost a right angle with the sphenoidal fissure. On the outer wall

of the orbit, just behind the external angular process of the
frontal bone, there is a depression for the **lachrymal gland.**
In the floor of the orbit the posterior opening of the **infraorbital canal,** for the passage of the superior maxillary division
of the trifacial nerve and the infraorbital artery, is to be seen.
There are to be seen, on the inner wall of the orbit, the canal
for the **nasal duct;** the **anterior ethmoidal foramen,** for
the passage of the anterior ethmoidal vessels and the nasal
nerve; and the **posterior ethmoidal foramen,** for the passage of the posterior ethmoidal vessels. On the inner wall there
is also a depression for the pulley of the superior oblique muscle
of the eyeball. (Morris, p. 96; Gray, p. 217.)

The base of the orbit is guarded by two folds of integu-
mentary and connective tissue which are spoken of as the **superior** and the **inferior eyelids.** The cleft between the eyelids is
called the **palpebral fissure.** The lids meet at either extremity of the palpebral fissure to form the **inner canthus**
and the **outer canthus.** The lids at the outer canthus form
an acute angle, while at the inner canthus the angle is more
obtuse and for a short distance the margins of the two lids lie
parallel.

The lids are attached, externally, to the malar bone by the
external tarsal ligament, and, internally, to the nasal process
of the superior maxillary bone by the **internal tarsal ligament** or **tendo oculi.** The internal tarsal ligament divides
into two processes one of which passes to the upper, and the
other to the lower lid. Between these two processes, at the
inner canthus of the eye, we find an isolated island of skin
called the **lachrymal caruncle.**

Each lid is composed of (1) the *skin,* (2) the *superficial
fascia,* (3) the *orbicularis palpebrarum muscle,* (4) the *tarsal plate,*
(5) the *cilia,* (6) the *Meibomian glands* and the *glands of Moll,*
and (7) the *conjunctiva.*

The **tarsal plate** is a layer of dense fibrous tissue which
gives stiffness to the lids. The levator palpebræ superioris muscle
is inserted into the tarsal plate of the upper lid.

The **Meibomian glands** are modified sebaceous glands

which open upon the free surface of the lid. They secrete an oily substance which prevents the lachrymal secretion from flowing over the cheek.

The **glands of Moll** are modified sudoriferous glands.

The **conjunctiva** is the layer of modified skin which lines the eyelids and which is reflected from the lids to the globe of the eye, which it invests. That portion of the conjunctiva which lines the lids is known as the **palpebral conjunctiva;** while the portion which covers the eyeball is spoken of as the **ocular conjunctiva.** The position at which the conjunctiva is reflected from the lid to the eyeball is termed the **fornix.** The fold of conjunctiva seen at the inner canthus of the eye, formed by the passage of the membrane from the lachrymal caruncle to the eyeball, is known as the **plica semilunaris.** The space between the eyelid and the eyeball is known as the **conjunctival cul-de-sac.** The conjunctiva is improperly spoken of as mucous membrane.

The lachrymal apparatus consists of (1) the *lachrymal gland,* (2) the *lachrymal ducts,* (3) the *lachrymal puncta,* (4) the *lachrymal canaliculi,* (5) the *lachrymal sac,* and (6) the *nasal duct.*

The **lachymal gland** is a small racemose gland, which is situated partly on the outer wall and partly on the roof of the orbit, just within the external angular process of the frontal bone.

The **lachrymal ducts,** about twelve in number, empty into the outer portion of the upper conjunctival cul-de-sac. The lachrymal secretion then passes from above, downward and inward, to the inner canthus of the eye, where, between the parallel portions of the two lids, there is a small pocket lying near the lachrymal caruncle, which is known as the **lacus lachrymalis.**

There is a small prominence on the straight portion of the margin of each lid termed the **lachrymal papilla,** on the summit of which a small opening, the **lachrymal punctum,** is to be seen.

The **superior** and **inferior lachrymal canaliculi** begin at the lachrymal puncta and pass inward to empty into the **lachrymal sac.**

From the lachrymal sac, the **nasal duct** passes through the canal between the lachrymal bone and the superior maxillary bone, to empty into the inferior meatus of the nose.

The muscles contained in the orbit are: (1) the *superior rectus*, (2) the *inferior rectus*, (3) the *internal rectus*, (4) the *external rectus*, (5) the *superior oblique*, (6) the *inferior oblique*, and (7) the *levator palpebræ superioris*.

The **four recti muscles** arise by a common tendon from the posterior portion of the orbit, above and external to the optic foramen, and on the inner side of the sphenoidal fissure. The **external rectus** has an additional head from the outer margin of the sphenoidal fissure. They are inserted into the sclerotic coat, in front of the equator of the eyeball.

The **superior oblique muscle** arises from the tendon common to the recti muscles. It passes forward, to a depression on the internal angular process of the frontal bone. In this depression the tendon of the muscle is inclosed in a fibrous sheath from which it passes backward to be inserted into the sclerotic coat in front of the equator of the eyeball.

The **inferior oblique muscle** arises from the anterior portion of the floor of the orbit and passes backward to be inserted into the posterior portion of the sclerotic coat of the eyeball.

The insertion of the four recti and the superior oblique muscle into the sclerotic coat is marked by a raised band of connective tissue, known as the **ligament of Zinn.**

The **levator palpebræ superioris muscle** arises from the common tendon of the recti muscles and passes forward to be inserted into the tarsal plate of the upper lid, and into the conjunctiva; it also sends a slip to the tendon of the superior rectus muscle

The **fascia** found in the orbit is continuous with the periosteum covering the bones composing the orbit, or **periorbita.** This fascia is reflected onto the eyeball, invests the sclerotic coat, and is then reflected onto the fat contained in the orbit. In this manner a closed sac is formed which is called the **capsule of Tenon.** The space between the two layers of this capsule is known as the **space of Tenon**; it is lined by endothelial cells and contains lymph. The muscles, as they pass to be inserted into the sclerotic coat of the eyeball, are ensheathed by processes from the capsule of Tenon. Fascial expansions pass from the sheaths of the internal rectus and the external rectus muscles to be inserted into the outer and inner walls of the orbit; these expansions are known, respectively, as the **internal** and **external check ligaments.**

At the anterior portion of the orbit a fascial expansion passes from the periorbita to the superior tarsal plate. This has been termed the **septum orbitale.**

A fascial expansion which passes from the covering of the inferior rectus muscle to the tarsal plate of the lower lid is called the **suspensory ligament of the eyeball.** This band of tissue passes transversely across the orbit.

The **eyeball or globe** is a spherical structure, having an antero-posterior diameter, a vertical diameter, and a transverse diameter. The antero-posterior diameter is the longest, measuring 24.5 millimetres; the transverse diameter measures 24 millimetres; and the vertical diameter measures 23.5 millimetres.

The eyeball has **three coats;** an outer, fibrous coat, the *sclerotic;* a middle, vascular coat, the *choroid;* and an inner, nervous coat, the *retina.*

The **sclerotic** is composed of dense white fibrous connective tissue. It envelops the eyeball for its posterior five-sixths, and anteriorly it is completed by a delicate, transparent structure which is known as the **cornea.** The junction of the cornea with the sclerotic is spoken of as the **sclero-corneal junction** or the **limbus corneæ.** The optic nerve pierces the posterior aspect of the sclerotic, about 3 millimetres to the nasal side of the posterior pole of the eyeball. The portion of the sclerotic through which the optic nerve fibres pass is known as the **lamina cribrosa.** The ciliary arteries and nerves also pierce the posterior aspect of the sclerotic, forming a circle around the optic nerve which is spoken of as the **circle of Zinn.** The venæ vorticosæ pierce the sclerotic at about the equator of the ball.

In the sclerotic, just behind the sclero-corneal junction, there is a circular, venous channel known as the **canal of Schlemm.** It empties into the anterior ciliary veins.

The **cornea** is composed of (1) the *epithelium*, (2) the *anterior limiting membrane*, (3) the *substantia propria*, (4) the *membrane of Descemet*, and (5) the *endothelium of Descemet.*

The **epithelium** is the continuation of the epithelium covering the conjunctiva; it is of the stratified squamous variety.

The **anterior limiting membrane** or **membrane of Bowman,** is the basement membrane which supports the epithelium.

The **substantia propria** is composed of parallel bundles of dense connective tissue fibres, which contain intercommunicating lacunæ between them. The lacunæ contain large connective tissue cells which are called the **corneal corpuscles.** The substantia propria contains no blood vessels.

The **membrane of Descemet** limits the posterior aspect of the substantia propria and is prolonged, laterally, into the substance of the iris, forming the **ligamentum pectinatum iridis.**

The **endothelium of Descemet** is a single layer of polyhedral cells which covers the posterior surface of the membrane of Descemet. (Piersol, p. 336; Morris, p. 861; Gray, p. 891.)

The **choroid** is the vascular coat of the eyeball. It lies beneath the sclerotic, covering the posterior five-sixths of the eyeball, and separated from it by the **subscleral lymph space.** The choroid is composed of (1) the *lamina suprachoroidea,* (2) the layer of *choroidal stroma,* (3) the *choriocapillaris,* and (4) the *vitreous lamina.*

The **lamina suprachoroidea** is the most external layer of the choroid; it contains no blood vessels.

The layer of **choroidal stroma** is a layer of connective tissue which supports a number of large blood vessels. It also contains numerous pigment cells.

The **choriocapillaris** contains a dense plexus of capillary blood vessels which serve for the supply of the underlying portions of the retina, as well as for the supply of the choroid.

The **vitreous lamina** separates the choroid from the underlying retina.

The **blood vessels of the choroid** are derived from branches of the short ciliary arteries and their corresponding veins. The arteries pass to the choriocapillaris where they break up into the complex capillary net work found in that situation. The veins, returning, are arranged in the form of whorls in the four poles of the eyeball, from which large vessels pass, the **venæ vorticosæ,** to pierce the sclerotic about half way between the limbus corneæ and the optic nerve.

At the anterior portion of its extent we may observe that the choroid becomes thickened, presents numerous fringe-like processes, and contains muscular tissue. This is the **ciliary region** of the choroid. It is composed of (1) the *ciliary ring*, (2) the *ciliary processes*, and (3) the *ciliary muscle.*

The **ciliary ring** gives origin to the ciliary processes.

The **ciliary processes,** about seventy in number, arise from the ciliary ring and project inward, into the posterior chamber of the eye. They are well supplied with blood vessels which are derived from the anterior and the long ciliary arteries. From these vessels the aqueous humor of the eye is filtered.

The **ciliary muscle** arises from the sclero-corneal junction and from the ligamentum pectinatum iridis. It passes backward to be inserted into the ciliary ring and into the ciliary processes. This muscle is of the involuntary type and is composed of three parts; the **radial fibres** pass toward the iris and are inserted into the **circular fibres,** which surround the base of the iris. The **meridional fibres** are inserted into the choroid, opposite to the ciliary processes. When the ciliary muscle contracts it pulls the choroid and the ciliary processes forward and inward. In this manner the suspensory ligament of the lens is relaxed and the lens is allowed to bulge forward.

The anterior one-sixth of the choroid is composed of a muscular curtain which is known as the **iris.** The iris is separated from the cornea by the anterior chamber of the eye. It is perforated by an opening known as the **pupil.** The ligamentum pectinatum iridis attaches the iris to the cornea. Posteriorly, the iris is continuous with the anterior portion of the choroid, in front of the ciliary processes. The iris contains a circular band of involuntary muscle, **sphincter pupillæ**, and radiating bundles of muscular tissue, **dilator pupillæ.**

The iris is supplied by branches of the long ciliary arteries. These vessels form an anastomosis at the base of the iris, **circulus major,** from which branches come off, which pass inward to the pupil to form a second anastomosis, **circulus minor,** around it. (Piersol, p. 342; Morris, p. 862; Gray, p. 894.)

The **retina** is the nervous coat of the eye; it is divided into the *pars optica,* the *pars ciliaris,* and the *pars iridica.* The **pars optica** of the retina is the innermost coat of the posterior five-sixths of the eyeball. At its posterior aspect, about 3 millimetres to the nasal side of the posterior pole of the eye, we see the **blind spot** or **optic disc,** which indicates the position at which the optic nerve leaves the retina to pass through the sclerotic and the choroid. At the posterior pole of the eye we find the **macula lutea,** in the center of which is a small depression, the **fovea centralis.** The macula lutea is the point of most acute vision.

The retina is composed of (1) the *pigment layer,* (2) the *layer of neuro-epithelium,* and (3) the *cerebral layer.*

The **pigment layer** is composed of highly pigmented, polyhedral, epithelial cells. It lies against the choroid. The cells composing the pigment layer send protoplasmic processes, loaded with pigment, down between the rods and cones.

The **layer of neuro-epithelium** is the receptive layer of the retina; it lies next to the pigment layer and is composed of the visual cells and of the sustentacular cells. The neuro-epithelial elements are known as the **rods** and the **cones.** These structures point *outward* and lie embedded in the pigment layer of the retina. In the greater part of the retina there are three or four rods to one cone. In the macula lutea, however, there are fully as many cones as rods; while in the fovea centralis the rods are entirly wanting. The rods contain the **visual purple** in their outer segments.

The **cerebral layer** of the retina contains the various cells and fibres concerned in transmitting visual impulses to the brain. The neuro-epithelial elements send processes into the cerebral layer which end in relation with the dendrits from cells in the outer portion of that layer. These cells are bipolar nerve cells and are known as the **rod bipolars** and the **cone bipolars.** The neurits from these cells pass inward to form terminal arborizations in relation with arborizations of the dendrits from the ganglion cells, which are situated in the deeper portion of the cerebral layer. The neurits of the ganglion cells pass into the optic nerves and thence to the brain. The **optic nerves,** strictly regarded, are the neurits of the rod bipolars and the cone bipolars (see page 55).

At the junction of the anterior one-sixth with the posterior five-sixths of the extent of the eyeball the retina becomes much thinner and the nervous tissue entirely disappears. This change in the thickness of the retina makes a distinct ring around the entire circumference of the eyeball, which is known as the **ora serrata.** From the ora serrata the thinned out retina, composed principally of pigmented cells, is prolonged forward over the inner surface of the ciliary body, **pars ciliaris,** and over the inner aspect of the iris, **pars iridica,** to terminate at the pupil.

The retina is supplied with nutrition from two sources. The **arteria centralis retinæ,** which is a branch of the ophthalmic artery, enters the optic nerve from its ventral surface, a short distance before the nerve leaves the sclerotic. It enters the retina at the optic disc and breaks up into branches which supply its cerebral layer. The pigment layer and the layer of neuro-epithelium are supplied by the choriocapillaris of the choroid. (Piersol, p. 351; Morris, p. 864; Gray, p. 898.)

There are three **chambers in the eye;** (1) the *anterior chamber,* (2) the *posterior chamber,* and (3) the *vitreous chamber.*

The **anterior chamber** of the eye is bounded, in front, by the cornea; and behind, by the anterior surface of the iris and the anterior portion of the lens, which appears at the pupil. It contains the aqueous humor. The ligamentum pectinatum iridis passes from the cornea to the iris in the lateral recesses of this chamber. It contains numerous intercommunicating lymphatic clefts which are known as the **spaces of Fontana.**

The **posterior chamber** is bounded, in front, by the posterior aspect of the iris; and behind, by the anterior surface of the lens and its suspensory ligament. The posterior chamber communicates with the anterior chamber through the pupil. The ciliary processes project into the lateral recesses of the posterior chamber. It contains the aqueous humor.

The **aqueous humor** is a modified lymph which is poured out from the vessels in the ciliary processes into the posterior chamber. It then passes through the pupil into the anterior chamber and is carried off through the spaces of Fontana into the canal of Schlemm to reach, finally, the anterior ciliary veins.

The **vitreous chamber** lies behind the iris and in front of the retina. It contains the vitreous body.

The **vitreous body** is a gelatinous substance which is held in place by a capsule, the **hyaloid membrane.** Opposite the ora serrata the hyaloid membrane passes by numerous trabeculæ of connective tissue, to be inserted into the capsule of the lens, forming the **suspensory ligament of the lens.** The position at which the hyaloid membrane divides to form the suspensory ligament of the lens is known as the **zone of Zinn.** The anterior portion of the vitreous body, with which the lens comes in relation, has no covering of hyaloid membrane. The depression in the vitreous body for the reception of the lens is known as the **patellar fossa.** (Morris, p. 869; Gray, p. 903.)

The **lens** is a solid body, which is situated behind the iris and in front of the vitreous body. It is composed of hexagonal fibres, which are epithelial in origin. The **capsule** is the fibrous membrane which surrounds the lens. Posteriorly, the lens fibres are in direct contact with the capsule; but anteriorly they are separated from it by a layer of polygonal epithelium. The **suspensory ligament of the lens** is derived from the hyaloid membrane which surrounds the vitreous body.

When the eye is at rest the ciliary processes press against the suspensory ligament of the lens and keep it tense. When the eye endeavors to appreciate near objects, the ciliary muscle contracts and pulls the ciliary processes and the suspensory ligament of the lens forward, thus permitting the lens to become more convex by its own elasticity. This process is called **accommo=dation.** (Morris, p. 864; Gray, p. 904.)

THE DEVELOPMENT OF THE EYE.

The first indication of the eye in the fetus appears, at about the fifteenth day, as an outgrowth from that portion of the anterior primary cerebral vesicle which subsequently becomes the interbrain, known as the **primary optic vesicle.** This vesicle grows out to the surface of the body and lies just beneath the ectoderm. At the position of junction of the primary optic vesicle with the surface ectoderm, the latter layer undergoes thickening to form the **lens pit.** The lens grows inward, becoming, as it grows, a vesicle, the **lens sac.** The growth of the lens sac invaginates the anterior wall of the primary optic vesicle, so that it lies immediately in front of the posterior wall. In this way the cavity of the primary optic vesicle becomes obliterated and a new vesicle is formed, between the primitive lens and the anterior portion of the primary optic vesicle, which is known as the **secondary optic vesicle or optic cup.** It will be seen that the wall of the primary optic vesicle is derived from the ectoderm, since it is an evagination from one of the cerebral vesicles.

The wall of the secondary optic vesicle is composed of two layers; first, a layer formed from the posterior wall of the primary optic vesicle; and second, a layer derived from the invaginated anterior wall of the primary optic vesicle. The **pigment layer of the retina** is derived from the posterior wall of the primary optic vesicle; the **neuro-epithelial layer** and the **cerebral layer of the retina** are derived from the invaginated anterior wall of the primary optic vesicle.

The **lens** is derived from the ectodermic vesicle which, by growing inward, invaginates the primary optic vesicle.

The invagination of the primary optic vesicle produces a fissure on the ventral aspect of the developing eye, which is spoken of as the **choroid fissure.** This fissure permits the mesoderm to pass into the eye between the lens and the secondary optic vesicle.

The **choroid, iris, sclerotic** and **cornea** are developed from the mesoderm which surrounds the secondary optic vesicle.

The **vitreous body** is formed from the mesoderm which grows into the eye through the choroid fissure.

The **conjunctiva** is formed from the ectoderm covering the surface of the embryo.

The **eyelids** are developed as folds of ectoderm containing mesodermic tissue.

The **lachrymal gland** is developed from epithelial plugs from the primitive conjunctiva.

The **nasal duct** is formed from the fissure between the lateral process and the superior maxillary process of the first visceral arch (see p. 15). (Quain, p. 83 ; A. T. O., p. 129.)

CHAPTER VII.

THE HEAD AND NECK.

The **superficial fascia** of the lateral cervical region is continuous with the superficial fascia of the thorax, the arm, and the back. It is composed of a layer of areolar tissue containing fat. In it the *platysma myoides muscle,* the *external jugular vein,* the *anterior jugular vein,* the *posterior jugular vein,* and *branches of the superficial cervical plexus* may be found.

The **platysma myoides muscle** is a broad sheet of muscular tissue which is contained in the superficial fascia of the neck. It **arises from** the pectoral fascia. It is **inserted into** the lower border of the jaw and into the skin at the angle of the mouth. It is **supplied by** the inframaxillary branch of the facial nerve. (Morris, p. 447; Gray, p. 407.)

The **external jugular vein** is formed by the union of the temporo-maxillary and the posterior auricular veins. It passes between the layers of the superficial fascia, beneath the platysma myoides muscle, in a line drawn from the angle of the jaw to the middle of the clavicle, to empty into the subclavian vein. The posterior jugular vein, the anterior jugular vein, the suprascapular vein, and the transversalis colli vein empty into it. It also receives a branch of communication from the internal jugular vein.

The **posterior jugular vein** begins at the confluence of several small venous twigs, just below the occipital bone. It empties into the external jugular vein.

The **anterior jugular vein** begins beneath the chin and empties into the external jugular vein. (Morris, p. 641; Gray, p. 653.)

The **deep fascia** of the neck is a robust layer of connective tissue which forms an investing sheath for the structures in that region. It begins at the ligamentum nuchæ and passes forward, as a single layer, until it reaches the posterior border of the trapezius muscle. It then divides into two layers, one of which passes in front of, and the other behind that muscle. At the anterior border of the trapezius muscle the two layers unite and pass forward, across the posterior triangle of the neck, until it reaches the posterior border of the sterno-mastoid muscle. Here it again divides into two layers which pass, one in front of, and the other behind the muscle. At the anterior border of the sterno-mastoid muscle the two layers of the fascia unite and pass, in a single layer, across the anterior triangle of the neck to join in the median line of the neck, from the chin to the sternum, with a similar layer from the opposite side. This latter portion, the **anterior layer of the deep cervical fascia,** is sometimes called the **cravat fascia.** At the lower portion of the neck, just above the sternum, the anterior layer of the deep cervical fascia splits into an anterior layer and a posterior layer. The anterior layer passes to be attached to the anterior surface of the first piece of the sternum; while the posterior layer passes backward and is attached to the posterior surface of the first piece of the sternum. Between these two divisions of the anterior layer of the deep cervical fascia there is a small, triangular space, known as the **space of Burns.** This space is occupied by a lymphatic gland, a small amount of fat, and the sternal head of the sterno-mastoid muscle.

Processes are given off from the anterior layer of the deep cervical fascia which pass upward to cover the parotid gland and the masseter muscle; they are known as the **parotid fascia** and the **masseteric fascia,** respectively.

The layer of the deep cervical fascia which lies behind the sterno-mastoid muscle may be called the **posterior layer of the deep cervical fascia.** From this portion of the deep fascia two processes come off. One of these passes forward and lies in front of the trachea. This is known as the **pretracheal fascia.** It is prolonged into the thorax to form the fibrous layer of the pericardium. The other process also passes forward; but in a plane deeper than the former, and lies in front of the bodies of the vertebræ and of the muscles attached to these bones. This process is known as the **prevertebral**

fascia. It is prolonged down into the thorax in front of the vertebræ and blends with the fascial lining of the posterior mediastinum.

The pretracheal fascia and the prevertebral fascia send off processes which unite to form the **sheath of the carotid blood vessels.**

Between the pretracheal and prevertebral fascias, a visceral compartment is formed, and behind the prevertebral fascia a muscular compartment is produced. The muscular compartment contains the muscles which are attached to the anterior aspect of the vertebræ. The visceral compartment contains the pharynx, the esophagus, the larynx, the trachea, the thyroid body, and the recurrent laryngeal nerves. Between the posterior wall of the pharynx and the prevertebral fascia there is a space, occupied by areolar tissue, which is known as the **postpharyngeal space.** There is a second muscular compartment between the pretracheal fascia and the anterior layer of the deep cervical fascia. This compartment contains the sterno-hyoid, the omo-hyoid, and the sterno-thyroid muscles.

The deep cervical fascia is prolonged downward, over the subclavian vessels, to help form the sheath of the axillary vessels.

Processes are given off from the deep cervical fascia which bind the central tendon of the omo-hyoid muscle to the first rib and the central tendon of the digastric muscle to the hyoid bone. (Morris, p. 466; Gray, p. 407.)

After the deep fascia is removed from the neck a quadrangular space may be defined which is bounded, *in front*, by the median line of the neck; *behind*, by the anterior border of the trapezius muscle; *below*, by the clavicle; and *above*, by the lower border of the inferior maxillary bone and a line drawn from the angle of the inferior maxillary bone to the mastoid process of the temporal bone.

The sterno-mastoid muscle passes across the quadrangle, diagonally, from its antero-inferior angle to its postero-superior angle, forming an anterior triangle and a posterior triangle. The anterior belly of the omo-hyoid muscle and the posterior belly of the digastric muscle pass across the anterior triangle, on their way to be attached to the hyoid bone, dividing it into three smaller triangles; the inferior carotid triangle, the superior carotid triangle, and the submaxillary triangle. The posterior belly of the omo-hyoid muscle passes across the posterior triangle, dividing it into two smaller triangles; the occipital triangle and the subclavian triangle.

The **occipital triangle** is bounded, *in front,* by the posterior border of the sterno-mastoid muscle; *behind,* by the anterior border of the trapezius muscle; and *below,* by the posterior belly of the omo-hyoid muscle. *The floor* is formed by the splenius capitis, the levator anguli scapulæ, the scalenus posticus, and the scalenus medius muscles. It contains the *spinal accessory nerve,* the *superficial cervical plexus,* the *transversalis colli artery and vein,* and the *post cervical lymphatics.* (Morris, p. 1106; Gray, p. 565.)

THE SPINAL NERVES.

There are **thirty-one pairs of nerves** which have their origins from the spinal cord; eight cervical, twelve dorsal, five lumbar, five sacral, and one coccygeal. These nerves are formed by the union of an anterior, motor root and a posterior, sensory root. The sensory root in each case bears a ganglion. The two roots pierce the dura mater of the spinal cord and leave the vertebral canal by passing through the intervertebral foramina. Just before they leave the intervertebral foramina the two roots unite to form a common trunk. The first cervical nerve passes through the foramen between the occiput and the atlas. This common trunk, divides into a large, anterior branch and a small, posterior branch. The posterior branches pass backward and supply the skin and the muscles of the back, by a cutaneous branch and a muscular branch. In the cervical, lumbar, and sacral regions the anterior divisions unite, in varying ways, to form certain plexuses. In the thoracic region, the anterior divisions of the dorsal nerves do not form a plexus; but, on the contrary, pass between the ribs as the intercostal nerves. These nerves in their course give off a lateral cutaneous branch, an anterior cutaneous branch, and muscular branches. (Morris, p. 803; Gray, p. 826.)

THE SUPERFICIAL CERVICAL PLEXUS.

The **superficial cervical plexus** is formed by the anterior divisions of the first, second, third, and fourth cervical nerves. It makes its appearance in the occipital triangle, opposite the middle of the posterior border of the sterno-mastoid muscle. It divides into a *superficial group* and a *deep group of branches*. The superficial branches pierce the deep cervical fascia, lie, for a short distance, between it and the superficial fascia, and finally, enter the superficial fascia to pass to the areas which they supply.

The **superficial branches** of the superficial cervical plexus are: (1) the *occipitalis minor*, (2) the *auricularis magnus*, (3) the *superficialis colli*, (4) the *suprasternal*, (5) the *supraacromial*, and (6) the *supraclavicular*.

The **deep branches** of the superficial cervical plexus are: (1) the *communicating*, (2) the *communicans hypoglossi*, (3) the *muscular*, and (4) the *phrenic*.

The **occipitalis minor nerve** passes along the posterior border of the sterno-mastoid muscle, to be distributed to the scalp in the occipital region, to the skin over the mastoid process of the temporal bone, and to the pinna.

The **auricularis magnus nerve** passes upward, across the sterno-mastoid muscle, to be distributed to the skin over the mastoid process, to the auricle, and to the skin over the parotid gland.

The **superficialis colli nerve** passes transversely inward and is distributed to the skin in the median line of the neck, from the chin to the sternum.

The **suprasternal**, the **supraclavicular**, and the **supraacromial nerves** pass downward, to be distributed to the skin over the sternum, below the clavicle, and over the acromion, respectively.

The **communicating branches** join the pneumogastric, the hypoglossal, and the sympathetic nerves.

The **communicans hypoglossi nerves** join with the descendens hypoglossi, in front of the carotid sheath, to form the ansa hypoglossi.

The **muscular branches** supply the prevertebral muscles.

The **phrenic nerve** is formed by branches from the third, fourth, and fifth cervical nerves. It passes downward, lying on the scalenus anticus muscle, between the subclavian artery and vein, behind the first rib, across the superior mediastinum, between the pleura and pericardium, to be distributed to the under surface of the diaphragm.

As the **right phrenic nerve** crosses the superior mediastinum, it lies to the outer side of the right innominate vein and of the superior vena cava. The **left phrenic nerve** lies to the left of the subclavian artery and of the arch of the aorta. The phrenic nerves give branches to the pleura and to the pericardium. (Morris, p. 809; Gray, p. 831.)

The **suboccipital nerve** is the posterior division of the first cervical nerve. It passes through the suboccipital triangle, supplies the muscles in that region, and sends a branch to the skin.

The **occipitalis major nerve** is the posterior division of the second cervical nerve. It pierces the complexus muscle and is finally distributed to the scalp in the occipital region. It also gives off muscular branches. (Morris, p. 806; Gray, p. 828.)

The **subclavian triangle** is bounded, *above*, by the posterior belly of the omo-hyoid muscle; *in front*, by the posterior border of the sterno-mastoid muscle; and *below*, by the clavicle. *The floor* is formed by the first rib and the first digitation of the serratus magnus muscle. It contains the *third portion of the subclavian artery*, the *brachial plexus of nerves*, the *suprascapular vessels*, the *transversalis colli vessels*, the *external jugular vein*, and the *apex of the lung*. (Morris, p. 1106; Gray, p. 565.)

THE SUBCLAVIAN ARTERY.

The **right subclavian artery** is a branch of the innominate artery. It passes from the sterno-clavicular articulation, in a curved manner, across the root of the neck, to the anterior border of the first rib, where it becomes the axillary artery. The subclavian artery is divided into three portions by the scalenus anticus muscle. The first portion extends from the sterno-clavicular articulation to the inner border of the scalenus anticus muscle; the second portion lies behind the scalenus anticus muscle; and the third portion extends from the outer border of the scalenus anticus muscle to the anterior border of the first rib.

RELATIONS.—The first portion of the vessel is crossed by the right innominate vein, the right internal jugular vein, the right vertebral vein, the right pneumogastic nerve, and the right phrenic nerve. The apex of the lung lies below it and the recurrent laryngeal nerve winds around it from before backward. In the second portion of its course, the artery lies behind the scalenus anticus muscle and above the apex of the lung. In the third portion of its course the brachial plexus lies above it and to the outer side. The external jugular, the suprascapular, and the transversalis colli veins, and the suprascapular artery lie in front of it. This portion of the vessel is found in the subclavian triangle. The subclavian vein lies below the subclavian artery and is separated from it by the scalenus anticus muscle.

The **left subclavian artery** is a branch of the arch of the aorta. It passes obliquely upward and outward, across the superior mediastinum, to reach the neck, just external to the left sterno-clavicular articulation. It then passes behind the scalenus anticus muscle, through the subclavian triangle, to the anterior border of the first rib, where it becomes the axillary artery. The first portion of the left subclavian artery extends from the arch of the aorta to the inner margin of the scalenus anticus muscle. The second and third portions have the same limits as the corresponding portions of the right vessel.

RELATIONS.—As the first portion of the left subclavian artery passes across the superior mediastinum the phrenic nerve crosses it and lies to the left, while the pneumogastric nerve passes to the right, separating it from the left common carotid artery. It is crossed by the left internal jugular, the left vertebral, and the left subclavian veins; and lower down, by the left innominate vein. It is overlapped by the left pleura and the an-

terior margin of the left lung. The thoracic duct arches over it
at the root of the neck. The relations of the second and third
parts of the left subclavian artery are the same as the relations
of the corresponding parts of the right artery.

The **branches** of the subclavian artery are: (1) the *thyroid
axis*, (2) the *vertebral*, (3) the *internal mammary*, and (4) the
superior intercostal.

The **thyroid axis** is a short trunk which arises from the
subclavian artery and breaks up, almost immediately, into the
inferior thyroid, the *suprascapular*, and the *transversalis colli
arteries.*

The **inferior thyroid artery** passes upward and inward
to supply the thyroid body. As it passes to its point of distribution it lies behind the sheath of the carotid blood vessels
and is crossed by the sympathetic nerve. The middle cervical
ganglion of the sympathetic nerve usually rests upon it. The
branches of the inferior thyroid artery are: (1) the *ascending
cervical*, (2) the *muscular*, (3) the *esophageal*, (4) the *tracheal*, and
(5) the *inferior laryngeal.*

The **ascending cervical artery** passes upward in the
neck to anastomose with branches of the vertebral and of the
ascending pharyngeal arteries.

The **suprascapular artery** passes outward, lying in front
of the third portion of the subclavian artery. It passes over
the transverse ligament of the scapula and enters the supraspinous fossa; it then passes around the base of the spinous process
of the scapula and enters the infraspinous fossa. It anastomoses
with the posterior scapular artery and with the dorsalis scapulæ
artery.

The **transversalis colli artery** passes outward, across the
occipital triangle. It passes beneath the trapezius and divides
into the *posterior scapular artery* and the *superficial cervical artery.*

The **posterior scapular artery** passes downward, along
the vertebral border of the scapula, and anastomoses with the
suprascapular artery.

The **superficial cervical artery** passes up the neck and

anastomoses with the superficial branch of the arteria princeps cervicis.

The **vertebral artery** passes upward, between the longus colli and the scalenus anticus muscles, to enter the costo-transverse foramen in the sixth cervical vertebra. It then passes successively through the upper costo-transverse foramina. After leaving the first foramen, it passes through a groove on the transverse process of the atlas, through the suboccipital triangle, and around the lateral mass of the atlas. It then pierces the posterior occipito-atlantal ligament and enters the skull, by passing through the foramen magnum. Here the two vertebral arteries unite to form the basilar artery.

The **branches** of the vertebral artery are: (1) the *muscular* (2) the *lateral spinal,* (3) the *posterior meningeal,* (4) the *posterior, spinal,* (5) the *anterior spinal,* and (6) the *posterior cerebellar.*

The **lateral spinal arteries,** five or six in number, pass through the intervertebral foramina and supply the spinal cord and the cervical vertebræ.

The **posterior spinal artery** passes downward along the posterior aspect of the spinal cord, to which it furnishes nutriment.

The **anterior spinal arteries** of the two sides unite to form a common branch which passes downward, lying on the anterior aspect of the cord.

The **basilar artery** is formed by the union of the two vertebral arteries. It lies in a groove on the ventral surface of the pons Varolii. The branches of the basilar artery are: (1) the *pontine,* (2) the *anterior cerebellar,* (3) the *superior cerebellar,* (4) the *auditory,* and (5) the *posterior cerebral.*

The **auditory artery** passes through the internal auditory meatus and supplies the internal ear.

The **posterior cerebral arteries** help to supply the brain. They give off *ganglionic branches,* to the nuclei at the base of the brain; and *cortical branches* to the cerebral cortex.

The **internal mammary artery** is a branch of the subclavian artery. It passes behind the first costal cartilage and enters the thorax. In the thorax, the internal mammary artery rests upon the costal cartilages of the first six ribs, about one-half inch outside the margin of the sternum. In the sixth intercostal space it divides into its terminal branches.

The **branches** of the internal mammary artery are: (1) the *comes nervi phrenici,* (2) the *mediastinal,* (3) the *pericardiac,* (4) the *anterior intercostals,* (5) the *anterior perforating,* (6) the *musculo-phrenic,* and (7) the *superior epigastric.*

The **anterior intercostal arteries** are for the supply of the first five intercostal spaces. Usually two branches are given off for each space; one of which passes along the lower border of the upper rib, the other of which passes along the upper border of the lower rib. They anastomose with the intercostal branches of the thoracic aorta.

The **anterior perforating arteries,** five in number, are branches of the internal mammary artery. They pass through the first five intercostal spaces and supply the pectoralis major muscle and the skin over the sternum. In the female, the second, third, and fourth perforating arteries supply the mammary gland.

The **musculo-phrenic artery** is distributed to the superior surface of the diaphragm. It gives off *anterior intercostal branches,* which occupy positions in the lower intercostal spaces similar to those occupied by the vessels given off from the internal mammary artery. It also sends twigs to the oblique muscles of the abdomen.

The **superior epigastric artery** is a branch of the internal mammary artery. It is given off in the sixth intercostal space and passes between the seventh costal cartilage, the ensiform process of the sternum, and the diaphragm. It then enters the sheath of the rectus muscle and, in the substance of that muscle, anastomoses with the deep epigastric branch of the external iliac artery.

The **superior intercostal artery** is a branch of the subclavian artery as that vessel lies behind the scalenus anticus muscle. It arches over the apex of the pleura, passes over the neck of the first rib, and is distributed to the first and second intercostal spaces. It gives off the *deep cervical artery,* which passes up the neck to anastomose with the arteria princeps cervicis. (Morris, p. 527; Gray, p. 576.)

The **inferior carotid triangle** is bounded, *in front,* by the median line of the neck; *behind,* by the anterior border of the sterno-mastoid muscle; and *above,* by the anterior belly of the omo-hyoid muscle. *The floor* is formed by the longus colli, the scalenus anticus, and the rectus capitis anticus major muscles. It contains the *sterno-hyoid* and *sterno-thyroid* muscles, the *common carotid,* the *vertebral,* and the *inferior thyroid arteries,* the *internal jugular vein,* the *pneumogastric, sympathetic, recurrent laryngeal,* and *descendens hypoglossi nerves,* the *trachea,* and the *thyroid body.*

The **superior carotid triangle** is bounded, *behind,* by the anterior border of the sterno-mastoid muscle; *above,* by the posterior belly of the digastric muscle; and *below,* by the anterior belly of the omo-hyoid muscle. *The floor* is formed by the thyro-hyoid muscle, the hyo-glossus muscle, and the inferior constrictor and the middle constrictor muscles of the pharynx. It contains the *common carotid artery* and its bifurcation into the *internal carotid* and *external carotid arteries;* the *superior thyroid,* the *lingual,* the *facial,* the *occipital,* and the *ascending pharyngeal arteries;* the *internal jugular,* the *superior thyroid,* the *lingual,* the *facial,* and the *ascending pharyngeal veins;* the *descendens hypoglossi,* the *sympathetic,* the *pneumogastric,* the *hypoglossal,* the *superior laryngeal,* the *spinal accessory,* and the *external laryngeal nerves;* the *pharynx* and the *larynx.*

The **submaxillary triangle** is bounded, *above,* by the lower border of the inferior maxillary bone and an imaginary line drawn from the angle of the inferior maxillary bone to the mastoid process of the temporal bone; *below,* by the posterior belly of the digastric muscle; *in front,* by the anterior belly of the digastric muscle. *The floor* is formed by the mylo-hyoid and the hyo-glossus muscles. It contains the *submaxillary gland;* the *facial,* the *submental,* the *mylo-hyoid,* the *external carotid,* and the *internal carotid arteries;* the *facial* and the *internal jugular veins;* the *glosso-pharyngeal,* the *pneumogastric,* and the *mylo-hyoid nerves;* the *stylo-glossus*, and *stylo-pharyngeus muscles;* and the *stylo-maxillary ligament.* (Morris p. 1104; Gray p. 563.)

THE COMMON CAROTID ARTERY.

The **right common carotid artery** is a branch of the innominate. It passes up the neck in a line drawn from the sterno-clavicular articulation to a point midway between the angle of the inferior maxillary bone and the mastoid process of the temporal bone. Opposite the upper border of the thyroid cartilage, in the superior carotid triangle, it bifurcates into the external carotid and the internal carotid arteries.

RELATIONS.—The common carotid artery lies internal to the internal jugular vein and is contained with it in a sheath, formed by processes from the pretracheal and the prevertebral fascias. In this sheath the pneumogastric nerve is also found, lying between and behind the vessels. Anteriorly, the vessel is overlapped by the anterior border of the sterno-mastoid muscle, which is known as the muscle of reference for the common carotid artery. The descendens hypoglossi and the communicans hypoglossi nerves, forming the ansa hypoglossi, lie in front of the sheath of the vessel. In its upper portion, the facial, lingual, and superior thyroid veins cross in front of its sheath. Posteriorly, the sympathetic nerve, the cardiac branches of the sympathetic and pneumogastric nerves, and the inferior thyroid artery are to be found. Internally, the common carotid artery is in relation with the trachea, the esophagus, and the lateral mass of the thyroid body.

The **left common carotid artery** is a branch of the arch of the aorta. It passes through the superior mediastinum to reach the left sterno-clavicular articulation, and then takes the same course, and has the same relations as does the right common carotid artery.

RELATIONS.—In the superior mediastinum, it is in relation, in front, with the thymus gland and the left innominate vein; on the right, with the trachea and **the beginning of the innominate artery**; on the left, with the left pneumogastric nerve, which separates it from the left subclavian artery, and the edge of the pleura; behind, with the trachea, the esophagus, the thoracic duct, and the left recurrent laryngeal nerve. At the root of the neck the internal jugular vein lies in front of the left common carotid artery.

The common carotid artery gives off no branches in the neck. (Morris p. 496; Gray p. 547.)

THE EXTERNAL CAROTID ARTERY.

The **external carotid artery** is a branch of the common carotid, given off at the upper border of the thyroid cartilage, in the superior carotid triangle. It lies nearer the mid-line than does its companion vessel, the internal carotid. It passes through the superior carotid and the submaxillary triangles, and enters the substance of the parotid gland, through which it passes, to break up into its terminal branches opposite the neck of the condyle of the inferior maxillary bone.

RELATIONS.—In front, the artery is crossed by the posterior belly of the digastric muscle, the stylo-hyoid muscle, the hypoglossal nerve, the lingual vein, the facial vein, the temporo-maxillary vein, and branches of the facial nerve. Behind, it is separated from the internal carotid artery by the stylo-pharyngeus and the stylo-glossus muscles, the stylo-hyoid ligament, the glosso-pharyngeal nerve, the pharyngeal branch of the pneumogastric nerve, and a part of the parotid gland. Internally, it is in relation with the wall of the pharynx, the tonsil, and the ramus of the inferior maxillary bone.

The **branches** of the external carotid artery are: (1) the *superior thyroid*, (2) the *lingual*, (3) the *facial*, (4) the *occipital*, (5) the *posterior auricular*, (6) the *ascending pharyngeal*, (7) the *superficial temporal*, and (8) the *internal maxillary*.

The **superior thyroid artery** is a branch of the external carotid artery, in the superior carotid triangle. It passes inward and downward to supply the thyroid body. The **branches** of the superior thyroid artery are: (1) the *infrahyoid*, (2) the *superior laryngeal*, (3) the *sterno-mastoid*, and (4) the *crico-thyroid*.

The **superior laryngeal artery** enters the larynx by piercing the thyro-hyoid membrane, in company with the superior laryngeal nerve.

The **crico-thyroid artery** rests on the crico-thyroid membrane.

The **lingual artery** is a branch of the external carotid artery, in the superior carotid triangle. It passes inward, to the tip of the greater cornu of the hyoid bone, it then passes beneath the hyo-glossus muscle to be distributed to the tongue. The **branches** of the lingual artery are: (1) the *suprahyoid*, (2) the *sublingual*, to the sublingual gland, (3) the *dorsalis linguæ*, to the dorsum of the tongue, in the region of the circumvallate papillæ, and (4) the *ranine*, to the ventral surface of the tongue, as far as the tip.

The **facial artery** is a branch of the external carotid artery, in the superior carotid triangle. It passes obliquely upward and inward, through the submaxillary triangle, beneath the posterior belly of the digastric muscle, the stylo-hyoid muscle, and the hypoglossal nerve, through a groove in the submaxillary gland, to the groove on the lower border of the inferior maxillary bone, just in front of the masseter muscle. From the anterior border of the masseter muscle, it passes to the angle of the mouth; from the angle of the mouth to the ala of the nose; and from the ala of the nose to the inner canthus of the eye. At the anterior border of the masseter muscle, the facial vein lies behind the artery and, in the submaxillary triangle, the facial vein lies superficial to the submaxillary gland; while the facial artery passes beneath the gland. The **branches** of the facial artery are: (1) the *ascending palatine*, (2) the *tonsillar*, (3) the *submaxillary*, (4) the *submental*, (5) the *muscular*, (6) the *inferior labial*, (7) the *inferior coronary*, (8) the *superior coronary*, (9) the *lateralis nasi*, and (10) the *angular*.

The **ascending palatine artery** supplies the soft palate and the tonsil.

The **tonsillar artery** is distributed to the tonsil.

The **submaxillary arteries** supply the submaxillary gland.

The **submental artery** is distributed to the tissues beneath the chin.

The **inferior labial artery** passes in the tissues of the chin, about midway between the lower lip and the lower border of the inferior maxillary bone.

The **inferior coronary** and the **superior coronary arteries** pass in the tissues of the lower and the upper lips, respectively.   They lie between the mucous membrane and the orbicularis oris muscle.

The **lateralis nasi artery** is distributed to the wing of the nose.

The **angular artery** passes upward, along the side of the nose, to anastomose with the nasal branch of the ophthalmic artery.

The **occipital artery** is a branch of the external carotid artery, in the superior carotid triangle.   It passes obliquely backward and upward to a groove on the mastoid process of the temporal bone, behind the groove for the posterior belly of the digastric muscle, and is finally distributed to the scalp.   As the artery passes backward, the hypoglossal nerve hooks around it, from behind forward, and lies in front of it.   In the scalp, it is accompanied by the occipitalis major nerve.   The **branches** of the occipital artery are: (1) the *arteria princeps cervicis*, (2) the *sterno-mastoid*, (3) the *auricular*, (4) the *mastoid*, (5) the *meningeal*, (6) the *muscular*, and (7) the *terminal.*

The **arteria princeps cervicis** is given off by the occipital artery, just before that vessel reaches the groove on the mastoid portion of the temporal bone.   It passes down the neck and divides into a *superficial branch*, which anastomoses with the superficial cervical artery; and a *deep branch,* which anastomoses with the profunda cervicis artery.

The **posterior auricular artery** is a branch of the external carotid artery, in the substance of the parotid gland.   It passes between the external auditory meatus and the mastoid process of the temporal bone, to reach the scalp.   The **branches** of the posterior auricular artery are: (1) the *stylo-mastoid,* (2) the *auricular,* and (3) the *mastoid.*

The **ascending pharyngeal artery** is a branch of the external carotid artery, in the superior carotid triangle.   It passes up the neck, resting on the rectus capitis anticus major muscle. It supplies the pharynx and sends branches to the cerebral meninges and to the soft palate.

The **superficial temporal artery** is one of the terminal branches of the external carotid artery. It is given off in the substance of the parotid gland, opposite the neck of the con- dyle of the inferior maxillary bone. It passes upward, through the parotid gland, and above the root of the zygoma. It then pierces the deep fascia, in company with the auriculo-temporal nerve, and becomes an occupant of the superficial fascia of the scalp, to which it is distributed. The **branches** of the super- ficial temporal artery are: (1) the *transverse facial,* (2) the *middle temporal,* (3) the *anterior temporal,* and (4) the *posterior temporal.*

The **transverse facial artery** is given off from the super- ficial temporal artery in the substance of the parotid gland. It passes across the face, midway between the zygoma and Sten- son's duct.

The **anterior temporal artery** passes forward, in the superficial fascia of the scalp, to anastomose with the supraorbital and the frontal branches of the ophthalmic artery.

The **posterior temporal artery** passes backward, in the superficial fascia of the scalp, to anastomose with the posterior auricular artery and with the anterior branch of the occipital artery.

The **internal maxillary artery** is the other terminal branch of the external carotid artery. It is given off in the substance of the parotid gland, opposite the neck of the condyle of the inferior maxillary bone. It passes: first, between the internal lateral liga- ment and the neck of the condyle of the inferior maxillary bone; second, through the zygomatic fossa, between the external and the internal pterygoid muscles; and finally, through the pterygo- maxillary fissure, into the spheno-maxillary fossa, where it breaks up into its terminal branches. The **branches** of the internal maxillary artery are: (1) the *tympanic,* (2) the *middle meningeal,* (3) the *small meningeal,* (4) the *inferior dental,* (5) the *pterygoid,* (6) the *masseteric,* (7) the *deep temporal,* (8) the *buccal,* (9) the *alveolar,* (10) the *infraorbital,* (11) the *Vidian,* (12) the *pterygo-*

palatine, (13) the *naso-palatine*, and (14) the *descending palatine.*

The **tympanic artery** passes through the Glasserian fissure and is distributed to the middle ear.

The **middle meningeal artery** passes through the foramen spinosum, in the great wing of the sphenoid bone, and is distributed to the dura mater of the brain. Just before this vessel passes through the foramen spinosum, it is included between the two roots of the auriculo-temporal nerve. In its intracranial course it passes over the antero-inferior angle of the parietal bone (pterion). This point is situated about one and one-half inches behind, and one inch above the external angular process of the frontal bone.

The **small meningeal artery** is also distributed to the dura mater of the brain. It enters the skull by passing through the foramen ovale, in the great wing of the sphenoid bone.

The **inferior dental artery,** accompanied by the inferior dental nerve, enters the inferior dental canal, in the inferior maxillary bone, by passing through the inferior dental foramen. It is distributed to the lower teeth. Just before this vessel passes through the inferior dental foramen, it gives off the *mylo-hyoid artery*, which lies in the mylo-hyoid groove, in company with the mylo-hyoid nerve, and supplies the mylo-hyoid muscle. In the anterior portion of the inferior dental canal, the inferior dental artery divides into the *incisive artery*, for the supply of the incisor teeth, and the *mental artery*, which passes through the mental foramen, in company with the mental nerve, for the supply of the tissues of the chin.

The **pterygoid,** the **masseteric,** the **deep temporal,** and the **buccal arteries** are given off in the zygomatic fossa. They supply the muscles, the names of which they bear.

The **alveolar artery** is given off in the spheno-maxillary fossa. It passes backward through the pterygo-maxillary fissure. to its points of distribution. It passes over the tuberosity of the superior maxillary bone, sending *posterior dental branches,* through the foramina in that bone, to the upper molar and the bicuspid teeth. Other branches pass to the gums as *gingival branches.*

The **infraorbital artery** is frequently spoken of as the continuation of the internal maxillary. It passes through the sphenomaxillary fissure, in company with the superior maxillary division of the trifacial nerve, to enter the orbit. It enters the infraorbital canal, in the floor of the orbit, passes through it, and leaves it by passing through the infraorbital foramen, to the face. As the vessel passes through the infraorbital canal it gives off *anterior dental branches* which supply the canine and incisor teeth.

The **Vidian artery** passes backward, through the Vidian canal, in company with the Vidian nerve. It sends branches to the pharynx and to the Eustachian tube.

The **pterygo-palatine artery** passes backward, through the pterygo-palatine canal, in company with the pterygo-palatine nerve, to be distributed to the pharynx.

The **naso-palatine artery** passes through the spheno-palatine foramen to enter the nose. It then lies, in company with the naso-palatine nerve, in a groove on the vomer and anastomoses with the anterior palatine artery. It supplies the nasal mucous membrane.

The **descending palatine artery** passes downward, through the posterior palatine canal, in company with the anterior palatine nerve. It emerges on the roof of the mouth, just behind the last molar tooth. Here it divides into the *anterior palatine artery* and the *posterior palatine artery*. The **anterior palatine artery** passes forward, in a groove on the hard palate, passes through the foramen of Stenson, enters the nose, and anastomoses with the naso-palatine artery. The **posterior palatine artery** passes backward to supply the soft palate. (Morris p. 501; Gray p. 551.)

THE INTERNAL CAROTID ARTERY.

The **internal carotid artery** is a branch of the common carotid artery. It is given off in the superior carotid triangle, opposite the upper border of the thyroid cartilage. It lies first, external to the external carotid artery; but soon, passes behind that vessel, on its way up the neck. It takes a course that would be indicated by the upper part of a line drawn from the sterno-clavicular articulation to a point midway between the angle of the inferior maxillary bone and the mastoid process of the temporal bone. It then enters the carotid canal in the petrous portion of the temporal bone, through which it passes in a forward direction. On emerging from the carotid canal, it lies on the cartilage which fills in the middle lacerated foramen and bends sharply on itself to pass upward, through the walls of the cavernous sinus, to a point opposite the posterior clinoid process. It then bends again and, passing forward, divides into its terminal branches opposite the anterior clinoid process.

RELATIONS.—In the neck, the internal carotid artery is separated from the external carotid artery, which lies in front of it, by the stylo-glossus and the stylo-pharyngeus muscles, the stylo-hyoid ligament, the glosso-pharyngeal nerve, the pharyngeal branch of the pneumogastric nerve, and the parotid gland. The internal jugular vein and the pneumogastric nerve are to its outer side, the nerve gradually getting between and behind the two vessels. The sympathetic nerve, the glosso-pharyngeal nerve, and the hypoglossal nerve lie behind it and the pharynx, and the tonsil lie internal to it. As it passes through the carotid canal in the petrous portion of the temporal bone, it is separated from the tympanum by a thin plate of osseous tissue. In the skull, it is in relation with the cavernous sinus, the abducens nerve, and the carotid and cavernous plexuses of the sympathetic system.

The **branches** of the internal carotid artery are: (1) the *tympanic*, (2) the *arteriæ receptaculi*, (3) the *pituitary*, (4) the *meningeal*, (5) the *posterior communicating*, (6) the *anterior choroid*, (7) the *anterior cerebral*, (8) the *middle cerebral* and (9) the *ophthalmic*.

The **tympanic artery** enters the middle ear.

The **arteriæ receptaculi** are distributed to the walls of the cavernous sinus.

The **pituitary branches** supply the pituitary body.

The **meningeal branches** are distributed to the dura mater of the brain.

The **posterior communicating artery** passes backward and joins with the posterior cerebral artery to help form the circle of Willis (see p. 24).

The **anterior choroid artery** passes through the transverse fissure of the cerebrum, in the velum interpositum, and helps to form the choroid plexus.

The **anterior cerebral artery** passes forward, in the longitudinal fissure of the cerebrum, winds over the genu of the corpus callosum, and then passes backward, resting on the dorsal surface of that body. The two anterior cerebral arteries are connected by the **anterior communicating artery** before they pass over the genu of the corpus callosum. In this manner the circle of Willis is completed anteriorly (see p. 24). The anterior cerebral arteries give off *ganglionic branches*, to the basal gray ganglia of the cerebrum, and *cortical branches*, to the cerebral cortex.

The **middle cerebral artery** passes outward, in the fissure of Sylvius, and gives off *ganglionic branches*, to the basal gray ganglia, and *cortical branches*, to the cerebral cortex. One of the ganglionic branches, larger than its fellows, is known as the **lenticulo-striate artery.** It passes through an opening in the anterior perforated space and supplies the external capsule, the lenticular nucleus, the internal capsule, and the caudate nucleus. It is frequently found ruptured in cases of cerebral apoplexy, and is, therefore, called the artery of cerebral hemorrhage.

The **ophthalmic artery** passes through the optic foramen, in company with the optic nerve. It enters the orbit, where it divides into numerous branches for the supply of that cavity and its contents. The **branches** of the ophthalmic artery are: (*a*) an orbital group, (1) the *lachrymal*, (2) the *supraorbital*, (3) the *muscular*, (4) the *anterior ethmoidal*, (5) the *posterior ethmoidal*, (6) the *palpebral*, (7) the *frontal* and (8) the *nasal;* and (*b*) an ocular group, (9) the *arteria centralis retinæ*, (10) the *long ciliary,* (11) the *short ciliary*, and (12) the *anterior ciliary.*

The **lachrymal artery** accompanies the lachrymal nerve and supplies the lachrymal gland. It gives off a recurrent branch which passes backward into the skull.

The **supraorbital artery** leaves the orbit, in company with the supraorbital nerve, by passing through the supraorbital foramen. It is distributed to the scalp and anastomoses with the anterior temporal artery.

The **muscular branches** supply the muscles which are contained in the orbit.

The **anterior ethmoidal artery** leaves the orbit, in company with the nasal nerve, by passing through the anterior ethmoidal foramen. It is distributed to the anterior ethmoidal cells and gives off branches to the dura mater of the brain.

The **posterior ethmoidal artery** passes through the posterior ethmoidal foramen and supplies the posterior ethmoidal cells. It, also, gives off branches to the dura mater of the brain.

The **palpebral branches** are distributed to the upper and the lower eyelids.

The **frontal artery** passes over the internal angular process of the frontal bone and is distributed to the scalp. It anastomoses with the anterior temporal artery.

The **nasal artery** leaves the orbit at the inner canthus of the eye. It anastomoses with the angular branch of the facial artery and is then distributed to the skin covering the nose.

The **arteria centralis retinæ** pierces the optic nerve on its ventral surface and enters the eye at the blind spot. It is distributed to the cerebral layer of the retina (see p. 101).

The **long ciliary arteries,** two in number, pierce the sclerotic and pass between that coat and the choroid, to supply the ciliary region (see p. 100).

The **short ciliary arteries,** five or six in number, also pierce the sclerotic, lying in a circle around the optic nerve, and are distributed to the choroid.

The **anterior ciliary arteries** are branches of the muscular twigs which are given off from the ophthalmic artery. They pierce the sclerotic, just behind the sclero-corneal junction and supply the ciliary region and the iris. (Morris p. 519; Gray p. 565.)

The **internal jugular vein** is formed by the union of the lateral sinus and the inferior petrosal sinus, after they have left the skull by passing through the jugular foramen. At the beginning of the internal jugular vein there is an ampulla, which occupies the jugular fossa on the petrous portion of the temporal bone. The vein then passes down the neck, so that a line

drawn from a point midway between the angle of the inferior
maxillary bone and the mastoid process of the temporal bone to
the sterno-clavicular articulation would represent its course. As
it passes down the neck, it lies to the outer side of the internal
carotid artery, and subsequently, to the outer side of the common carotid artery. It is contained in the same sheath as are
these vessels, the pneumogastric nerve passing between them
and behind them. At the sterno-clavicular articulation, the internal jugular vein joins with the subclavian vein to form the innominate vein. The thoracic duct empties into the point of union
of the internal jugular and the subclavian veins, on the left side.
The right lymphatic duct empties into the corresponding point
on the right side. The internal jugular vein receives the pharyngeal, the lingual, the facial, and the superior and the middle
thyroid veins. It also receives a communicating branch from
the external jugular vein. (Morris p. 651; Gray p. 654.)

The **facial vein** begins at the inner canthus of the eye,
where it communicates with the nasal vein. It passes directly
across the face from the inner canthus of the eye to the anterior
border of the masseter muscle. It then passes above the submaxillary gland to empty into the internal jugular vein. It sends
a communicating branch to the external jugular vein. At the anterior border of the masseter muscle, the facial vein lies posterior
to the facial artery. In its course across the face, the facial vein
lies beneath the zygomaticus major muscle. (Morris p. 637; Gray
p. 652.)

THE SCALP.

The scalp is composed of (1) the *skin*, (2) the *superficial
fascia*, (3) the *occipito-frontalis aponeurosis*, (4) the *areolar tissue*,
and (5) the *pericranium*. In the temporal region we find two
additional layers; the *temporal fascia* and the *temporal muscle*.
These structures lie between the areolar tissue and the pericranium.

The blood-vessels and nerves of the scalp run in the superficial fascia. The scalp is supplied by the following **arteries:** the
frontal and supraorbital, branches of the ophthalmic; the anterior

temporal and the posterior temporal, branches of the superficial temporal; the posterior auricular and the occipital, branches of the external carotid.

The following **sensory nerves** are distributed to the scalp: the supratrochlear and the supraorbital, branches of the ophthalmic division of the trifacial; the temporal branch of the temporomalar, a branch of the superior maxillary division of the trifacial; the temporal branch of the auriculo-temporal, a branch of the inferior maxillary division of the trifacial; the auricularis magnus and the occipitalis minor, branches of the superficial cervical plexus; the occipitalis major,[1] the posterior division of the second cervical; and the suboccipital,[1] the posterior division of the first cervical nerve.

The **motor nerves** which are found in the scalp are as follows; the temporal branch and the posterior auricular branch of the facial. The suboccipital and the occipitalis major nerves contain motor fibres.

---

[1] These are mixed nerves.

CHAPTER VIII.

THE UPPER EXTREMITY.

The **superficial fascias** of the pectoral region and of the arm are continuous with each other and with the superficial fascia of the neck. In the forearm the superficial veins contained in the superficial fascia are four in number.

The **superficial radial vein** is situated on the outer border of the forearm. The **superficial median vein** is found in the middle of the forearm. The **superficial anterior** and the **superficial posterior ulnar veins** are found on the anterior and posterior aspects, respectively, of the inner side of the forearm. These veins begin in plexuses situated in the hand. Just below the bend of the elbow, the superficial median vein bifurcates into the *median cephalic vein* and the *median basilic vein*. The **median cephalic vein** passes outward and joins with the superficial radial vein to form the cephalic vein. The **median basilic vein** passes inward to join with the common superficial ulnar vein, formed by the union of the anterior and posterior superficial ulnar veins, to form the basilic vein. The **cephalic vein** passes up the outer side of the arm, in the groove in the superficial fascia made by the outer border of the biceps muscle; lies, in company with the descending branch of the acromio-thoracic artery, in the groove between the deltoid and the pectoralis major muscles, and finally, pierces the costocoracoid membrane to empty into the axillary vein. The **basilic vein** passes up the inner side of the arm, in the groove in the superficial fascia made by the inner border of the biceps muscle, and, at the junction of the middle and lower thirds of the arm, pierces the deep fascia, in company with the internal cutaneous nerve, to lie deeply placed, in relation with the brachial artery. It finally joins with the venæ comites of the brachial artery to form the axillary vein. (Morris p. 664; Gray p. 662.)

Between the two layers of the superficial fascia in the pectoral region, the mammary gland is to be found. The **mammary gland** is a modified sebaceous gland of the compound racemose type. There is much fat between the lobules of which the gland is composed and processes of the superficial fascia dip down into the organ forming its **suspensory ligaments.** The ducts from the acini in the various lobules converge and open, as the lactiferous ducts, on the free surface of the nipple. Each lactiferous duct presents a dilation of its lumen just before it terminates, which is known as the **ampulla.**

The nipple contains some erectile tissue; but is entirely devoid of fat. (Morris, p. 1084; Gray, p. 1178.)

The **deep fascia** of the thorax, which covers the pectoralis major muscle, is known as the **pectoral fascia.** The **clavipectoral fascia** or **costo-coracoid membrane** is found beneath the pectoralis major muscle. It begins, above, at the clavicle where it is attached on either side of the subclavius muscle. It then passes downward, to the upper border of the pectoralis minor muscle, extending laterally from the coracoid process of the scapula to the costal cartilage of the first rib. This quadrangular sheet of tissue is the true costo-coracoid membrane. At the upper border of the pectoralis minor muscle, the clavi-pectoral fascia divides and encloses the muscle, uniting again to form a single layer at its lower border. The clavi-pectoral fascia is then joined by the pectoral fascia, which comes around the lower border of the pectoralis major muscle, and the two pass across the axilla as the **axillary fascia,** forming the floor of that space and blending with the deep fascia of the arm. The costo-coracoid membrane is **pierced by** the cephalic vein, the acromiothoracic artery and vein, and the external anterior thoracic nerve. That portion of the clavi-pectoral fascia which lies behind the pectoralis minor muscle is intimately united to the underlying sheath of the axillary blood vessels. This sheath is derived from the deep cervical fascia (see p. 107).

In the arm, the deep fascia sends processes down to be attached to the external and internal supracondyloid ridges of the humerus. These processes are known, respectively, as the

external and **internal intermuscular septa.** They, with the humerus, divide the arm into an anterior muscular compartment and a posterior muscular compartment. (Morris, p. 331; Gray, p. 466.)

The **bicipital fascia** is an aponeurotic slip which passes from the tendon of the biceps muscle, inward, to blend away into the deep fascia of the forearm.

The **anterior annular ligament** is a robust layer of fibrous tissue which is attached to the scaphoid and the trapezium, on the radial side, and to the pisiform and the unciform process of the unciform, on the ulnar side of the wrist. Beneath the anterior annular ligament and above the carpal bones we find the four tendons of the *flexor sublimis digitorum muscle,* the four tendons of the *flexor profundus digitorum muscle,* the tendon of the *flexor longus pollicis muscle,* and the *median nerve* passing into the palm of the hand. Above the anterior annular ligament the *ulnar artery,* the *ulnar nerve,* the *superficialis volæ* branch of the radial artery, and the tendon of the *palmaris longus muscle* pass into the palm of the hand. (Morris, p. 329; Gray, p. 489.)

The **posterior annular ligament** is less robust than the anterior annular ligament. It is attached to the lower end of the radius, externally, and to the pisiform and cuneiform bones, internally. Beneath this ligament there are six compartments, through which the extensor muscles of the wrist and fingers pass in the following order, from without inward; first, the *extensor ossis metacarpi pollicis* and the *extensor brevis pollicis;* second, the *extensor carpi radialis longior* and the *extensor carpi radialis brevior;* third, the *extensor longus pollicis;* fourth, the *extensor communis digitorum* and the *extensor indicis;* fifth, the *extensor minimi digiti;* and sixth, the *extensor carpi ulnaris.* (Morris, p. 329; Gray, p. 490.)

The **interosseous membrane** is a robust layer of fibrous tissue which stretches across the space between the radius and the ulna.

The **palmar fascia** is the deep fascia of the palm of the hand. The central portion is of triangular shape, the apex being directed upward. The central, triangular portion is quite robust and covers in the structures in the palm, protecting them from

pressure. At a position in the palm which about corresponds
with the first transverse furrow of the integument, the palmar fascia divides into four slips, which pass downward above the metacarpal bones. Opposite the metacarpo-phalangeal articulation, each
of these primary slips divides into two secondary slips, which
pass forward and are inserted into the anterior ligaments of the
metacarpo-phalangeal articulations. Between the primary divisions
of the palmar fascia, the digital arteries and the digital nerves pass
on their way to the fingers. Between the secondary slips, the
flexor tendons, enclosed in their synovial sheaths, pass to the
fingers. Lateral expansions of the palmar fascia, thinner than
the central portion, cover the thenar and the hypothenar emin-
ences. These fascias may be called the **thenar fascia** and the
hypothenar fascia. The tendon of the palmaris longus muscle
is inserted into the apex of the palmar fascia. (Morris, p. 351 ;
Gray, p. 490.)

THE AXILLA.

The **axilla** is bounded, *in front,* by the pectoralis major
muscle, the pectoralis minor muscle, and the costo-coracoid membrane ; *behind,* by the subscapularis muscle, the teres major muscle, and the latissimus dorsi muscle ; *internally,* by the first four
ribs, the first three external intercostal muscles, and the first five
digitations of the serratus magnus muscle ; *externally,* by the
humerus, the short head of the biceps muscle, and the coraco-
brachialis muscle. *The floor* is formed by the axillary fascia.
The apex corresponds to the interval between the clavicle and the
first rib. The axilla contains the *axillary artery and its branches,*
the *axillary vein and its tributaries,* the *brachial plexus of nerves
and its branches,* the *axillary lymphatics,* and *fat.*

On the posterior wall of the axilla the subscapularis muscle,
passing to its insertion into the lesser tuberosity of the humerus,
and the teres major muscle, passing to its attachment to the bottom of the bicipital groove, make with the humerus, a **triangular
space.** This triangle is crossed by the long head of the triceps
muscle and is, thus, converted into an outer, *quadrangular space,*
and an inner, *triangular space.* The **quadrangular space** is

bounded, *externally*, by the humerus; *internally*, by the long head of the triceps muscle; *above*, by the subscapularis* (teres minor) muscle; and *below*, by the teres major muscle. It transmits the *posterior circumflex artery* and the *circumflex nerve*. The **triangular space** is bounded, *externally*, by the long head of the triceps muscle; *above*, by the subscapularis (teres minor) muscle; and *below*, by the teres major muscle. It transmits the *dorsalis scapulæ artery*. The pectoralis minor muscle passing across the axilla, to be inserted into the coracoid process of the scapula divides that space into three parts. The first part is situated between the anterior border of the first rib and the upper border of the pectoralis minor muscle; the second part is situated behind the pectoralis minor muscle; and the third part extends from the lower border of the pectoralis minor muscle to the lower border of the teres major muscle. (Morris, p. 1160; Gray, p. 587.)

THE BRACHIAL PLEXUS.

The **brachial plexus** is formed by the anterior divisions of the fifth, sixth, seventh, and eighth cervical, and the first dorsal nerves. This plexus is formed in the neck and makes its appearance between the scalenus anticus and the scalenus medius muscles. It passes across the subclavian triangle, beneath the clavicle, and enters the axilla. The fifth and sixth cervical nerves unite, soon after they leave the intervertebral foramina, to form the **superior trunk** of the brachial plexus. The seventh cervical nerve forms the **middle trunk** of the brachial plexus. The eighth cervical and the first dorsal nerves unite to form the **inferior trunk** of the brachial plexus. Each of these three trunks soon divides into an anterior division and a posterior division. The posterior divisions of the three trunks unite to form the **posterior cord** of the brachial plexus. The anterior divisions of the superior and middle trunks unite to form the **outer cord** of the brachial plexus. The anterior division of the inferior trunk forms the **inner cord** of the brachial plexus.

---

* The subscapularis muscle forms the upper boundary of these spaces if they are studied from the anterior aspect. If they are viewed from the posterior surface, the upper boundary will be seen to be the teres minor muscle.

RELATIONS.—In the neck, the brachial plexus lies above and to the outer side of the subclavian artery. It is crossed by the transversalis colli and the suprascapular vessels. In the first portion of the axilla, the brachial plexus lies above and to the outer side of the axillary artery; in the second portion of the axilla, the brachial plexus surrounds the axillary artery; and in the third portion of the axilla, the branches of the brachial plexus surround the axillary artery.

The **branches** of the brachial plexus **above the clavicle** are: (1) the *posterior thoracic*, (2) the *suprascapular*, (3) the *muscular,* and (4) the *communicating.*

The **posterior thoracic nerve** or the **external respiratory nerve of Bell** is formed by branches from the fifth, sixth, and seventh cervical nerves. It is formed in the substance of the scalenus medius muscle and enters the axilla by passing across the first rib. In the axilla, it is the most posterior structure, lying behind the axillary artery and on the serratus magnus muscle, to which it is distributed.

The **suprascapular nerve** passes through the suprascapular notch, beneath the transverse ligament of the scapula. It enters the supraspinous fossa and sends branches to the supraspinatus muscle; it then passes around the base of the spine of the scapula and enters the infraspinous fossa. In the infraspinous fossa it gives branches to the infraspinatus muscle.

The **muscular branches** are for the supply of the rhomboideus major, the rhomboideus minor, the subclavius, the scalenus anticus, the scalenus medius, the scalenus posticus, and the longus colli muscles.

The **communicating branches** join the phrenic nerve.

The **branches of the outer cord** of the brachial plexus are: (1) the *external anterior thoracic*, (2) the *musculo-cutaneous,* and (3) the *outer head of the median.*

The **external anterior thoracic** nerve is a branch of the outer cord of the brachial plexus. It pierces the costo-coracoid membrane and is distributed to the pectoralis major muscle.

The **musculo-cutaneous nerve** is a branch of the outer cord of the brachial plexus. It pierces the coraco-brachialis muscle and then lies between the biceps and the brachialis anticus muscles. Just above the bend of the elbow, it pierces the deep fascia and lies external to the tendon of the biceps. It then divides into an anterior branch and a posterior branch, which are distributed to the skin on the radial side of the forearm. The anterior branch passes beneath the median cephalic vein. In the arm, it gives branches to the coraco-brachialis muscle, the biceps muscle, and the brachialis anticus muscle.

The **branches of the inner cord** of the brachial plexus are: (1) the *internal anterior thoracic,* (2) the *lesser internal cutaneous,* (3) the *internal cutaneous,* (4) the *ulnar,* and (5) the *inner head of the median.*

The **internal anterior thoracic nerve** is a branch of the inner cord of the brachial plexus. It passes between the axillary artery and the axillary vein and supplies the pectoralis minor muscle.

The **lesser internal cutaneous nerve** or **nerve of Wrisberg** is a branch of the inner cord of the brachial plexus. It forms a plexus on the inner aspect of the arm with the intercosto-humeral nerve. It supplies the skin on the inner side of the arm and ends over the olecranon.

[The **intercosto-humeral nerve** is the lateral cutaneous branch of the second intercostal nerve. It passes through the second intercostal space, across the floor of the axilla, to the inner aspect of the arm, where it forms a plexus with the lesser internal cutaneous nerve.] .

The **internal cutaneous nerve** is a branch of the inner cord of the brachial plexus. It passes down the arm, lying internal to the brachial artery. At the junction of the middle and lower thirds of the arm, it pierces the deep fascia, in company with the basilic vein. It divides into an anterior branch and a posterior branch, which are distributed to the skin on the inner side of the forearm. The anterior branch usually passes above the median basilic vein.

The **ulnar nerve** is a branch of the inner cord of the brachial plexus. It passes down the arm, lying internal to the axillary and brachial arteries. At the junction of the middle and lower thirds of the arm, it pierces the internal intermuscular septum, in company with the inferior profunda artery. It then lies between the internal condyle of the humerus and the olecranon, passes between the two heads of the flexor carpi ulnaris muscle, and enters the forearm. In the forearm, it lies, in company with the ulnar artery, which is placed to its outer side, on the flexor profundus digitorum muscle, and beneath the flexor carpi ulnaris muscle. Just above the wrist, the nerve gives off a posterior branch. The posterior branch passes to the posterior aspect of the forearm and supplies the dorsal surface of the fifth finger and one-half of the dorsal surface of the fourth finger. The ulnar nerve then passes, in company with the ulnar artery, over the anterior annular ligament, between the pisiform bone and the unciform process of the unciform bone, and supplies the anterior surface of the fifth finger and the adjacent half of the anterior surface of the fourth finger. It gives off a superficial palmar branch, which supplies the skin of the ulnar side of the palm of the hand, and a deep palmar branch, which accompanies the deep palmar arch. In its course, the ulnar nerve gives branches to the elbow and the wrist joints and to the following muscles: the flexor carpi ulnaris, the inner half of the flexor profundus digitorum, the abductor minimi digiti, the flexor brevis minimi digiti, the opponens minimi digiti, the palmaris brevis, the third and the fourth lumbricals, the palmar and the dorsal interossei, the abductor pollicis, and one head of the flexor brevis pollicis.

The **median nerve** is formed by a branch from the outer cord of the brachial plexus, the outer head of the median nerve, and a branch from the inner cord of the brachial plexus, the inner head of the median nerve. These two heads unite in front of the axillary artery and the trunk formed by their union lies, first, outside the brachial artery, then crosses above it, to pass down the arm on its inner side. It passes through the cubital fossa, in which it is the innermost structure, between the two heads of the pronator radii teres muscle, and enters the forearm. In the forearm, it has a median position, lying between the flexor sublimis digitorum and the flexor profundus digitorum muscles.

accompanied by the median artery. It passes beneath the anterior annular ligament of the wrist and enters the palm of the hand. It supplies the palmar surfaces of the thumb, the second, and the third fingers, and the outer half of the fourth finger. It sends branches to the dorsal aspects of the second, third, and fourth fingers. According to some authorities, the median nerve sends branches to the dorsal surfaces of the third phalanges of the thumb and of the little finger. In its course, just as it passes between the two heads of the pronator radii teres muscle, the median nerve gives off the *anterior interosseous nerve*. The **anterior interosseous nerve** passes down the forearm, resting on the anterior aspect of the interosseous membrane. It is accompanied by the anterior interosseous artery. It supplies the flexor longus pollicis, one-half of the flexor profundus digitorum, and the pronator quadratus muscles. The median nerve also supplies the flexor carpi radialis, the palmaris longus, the flexor sublimis digitorum, and the pronator radii teres muscles. It gives filaments to the elbow joint and cutaneous branches to the palm of the hand. In the hand, it supplies the first and second lumbricals, the opponens pollicis, the abductor pollicis, and one head of the flexor brevis pollicis muscles.

The **branches of the posterior cord** of the brachial plexus are: (1) the *three subscapular*, (2) the *circumflex*, and (3) the *musculo-spiral.*

The **three subscapular nerves** are branches of the posterior cord of the brachial plexus. They supply the subscapularis, the teres major, and the latissimus dorsi muscles.

The **circumflex nerve** is a branch of the posterior cord of the brachial plexus. It passes backward, through the quadrangular space at the back of the shoulder, in company with the posterior circumflex artery. It divides into an *anterior branch* and a *posterior branch*. The **anterior branch** passes between the deltoid muscle and the neck of the humerus, sending filaments into the muscle, some of which pass through the muscle to supply the skin. The **posterior branch** gives filaments to the deltoid and the teres minor muscles and then, passing from beneath the posterior border of the deltoid muscle, supplies the skin of the shoulder. The circumflex nerve sends a branch to the shoulder joint before it divides.

The **musculo-spiral nerve** is a branch of the posterior cord of the brachial plexus. It lies beneath the axillary and the brachial arteries, passes around the humerus in the musculo-spiral groove, accompanied by the superior profunda artery. As it lies in the musculo-spiral groove, it is placed between the internal and the external heads of the triceps muscle. At the junction of the middle and lower thirds of the arm it pierces the external intermuscular septum and lies between the brachialis anticus and the supinator longus muscles. Before it passes into the musculo-spiral groove, the musculo-spiral nerve gives off the **internal cutaneous branch,** which is distributed to the skin on the inner and posterior aspects of the arm. As it passes through the musculo-spiral groove, the nerve gives off the *superior* and the *inferior external cutaneous branches.* The **superior external cutaneous branch** supplies the skin on the anterior surface of the arm. The **inferior external cutaneous branch** passes downward to the posterior surface of the forearm, to the skin of which it is distributed, lying between the posterior branches of the musculo-cutaneous and the internal cutaneous nerves. The musculo-spiral nerve sends muscular branches to the triceps, the brachialis anticus, the supinator longus, the extensor carpi radialis longior, and the anconeus muscles. Between the brachialis anticus and the supinator longus muscles, the musculo-spiral nerve divides into the *posterior interosseous nerve* and the *radial nerve.*

The **posterior interosseous nerve** pierces the supinator brevis muscle and then lies between the superficial and deep layers of extensor muscles, on the posterior aspect of the forearm. It ends in a ganglion, which is situated on the posterior ligaments of the carpus. It supplies the extensor carpi radialis brevior, the extensor carpi ulnaris, the extensor communis digitorum, the extensor minimi digiti, the extensor indicis, the extensor ossis metacarpi pollicis, the extensor longus pollicis, and the extensor brevis pollicis muscles. The ganglion sends branches to the carpal joints.

The **radial nerve** passes downward, beneath the supinator longus muscle and above the flexor longus pollicis muscle. In the middle third of the forearm, it accompanies the radial artery, lying to its outer side. At the junction of the middle and lower thirds of the forearm, the radial nerve winds backward. It is distributed to the dorsal surfaces of the first, the second, the third fingers, and one-half of the fourth finger. (Morris, p. 813; Gray, p. 834.)

The **cutaneous nerve supply of the arm** is as follows: anterior surface, from within outward, the intercosto-humeral, the lesser internal cutaneous, the internal cutaneous branch, and the superior external cutaneous branch of the musculo-spiral. Posterior surface, from within outward, the intercosto-humeral and the lesser internal cutaneous, the internal cutaneous branch of the musculo-spiral, and the circumflex. Over the shoulder, the cutaneous nerves come from the supraacromial branches of the superficial cervical plexus and from the circumflex.

The **cutaneous nerve supply of the forearm** is as follows: on the anterior surface, the anterior branch of the musculo-cutaneous nerve to the radial side and the anterior branch of the internal cutaneous nerve to the ulnar side. On the posterior surface, the posterior branch of the musculo-cutaneous nerve to the radial side, the posterior branch of the internal cutaneous nerve to the ulnar side, and the inferior external cutaneous branch of the musculo-spiral nerve, between the other two branches.

The **cutaneous nerve supply of the hand** is as follows: on the palmar surface, the median nerve supplies the first, second and third fingers, and one-half the fourth finger. The ulnar nerve supplies the fifth finger and one-half the fourth finger. Each nerve sends a branch to the palm. On the dorsal surface, the radial nerve supplies the first, second, and third fingers, and one-half the fourth finger. The ulnar nerve supplies the fifth finger and one-half the fourth finger. Each nerve sends a branch to the dorsum of the hand. The median nerve sends branches to the dorsal surfaces of the second, third, and fourth fingers. According to some authorities the median nerve also sends branches to the dorsal surfaces of the first and fifth fingers.

THE AXILLARY ARTERY.

The **axillary artery** is the continuation of the subclavian artery. It begins at the anterior border of the first rib and passes through the axilla, lying along the outer wall of that space, and near its anterior boundary. The upper part of a line drawn from the middle of the clavicle to the middle of the bend of the elbow, when the arm is at right angles to the body, would represent its course. It ends at the lower border of the tendon of the teres major muscle.

RELATIONS.—In the first portion of the axilla, the brachial plexus lies above it and external to it. In the second portion of the axilla, the brachial plexus surrounds it. In the third part of the axilla, the branches of the brachial plexus surround it. The axillary vein lies internal to it throughout its course. It is covered by the pectoralis major and the pectoralis minor muscles and by the costo-coracoid membrane.

The **branches** of the axillary artery are: (1) the *superior thoracic*, (2) the *acromio-thoracic*, (3) the *long thoracic*, (4) the *alar thoracic*, (5) the *subscapular*, (6) the *anterior circumflex*, and (7) the *posterior circumflex*.

The **superior thoracic artery** passes along the upper border of the pectoralis minor muscle. It sends branches to the pectoralis major and the pectoralis minor muscles and to the thoracic wall.

The **acromio-thoracic artery** pierces the costo-coracoid membrane and divides into *thoracic, acromial,* and *descending branches*. The **thoracic branches** are distributed to the pectoralis major muscle. The **acromial branches** supply the deltoid muscle and help to form the crucial anastomosis. The **descending branch** passes, with the cephalic vein, in the groove between the pectoralis major and the deltoid muscles.

The **long thoracic artery** passes along the lower border of the pectoralis minor muscle and is distributed to the serratus magnus muscle and to the inner wall of the axilla.

The **alar thoracic artery** supplies the fat and the lymphatic glands which are found in the axilla.

The **subscapular artery** passes along the lower border of the subscapularis muscle and helps to form the scapular anastomosis, in addition to supplying the subscapularis muscle. The **dorsalis scapulæ artery** is a branch of the subscapular artery. It passes through the triangular space at the back of the shoulder (see p. 132) and helps to form the scapular anastomosis.

The **scapular anastomosis** is rich and is formed as follows: on the posterior border of the scapula, the posterior scapular artery anastomoses with the subscapular artery and with a branch from one of the intercostal arteries. On the axillary border of the scapula, we find the subscapular artery. In the supraspinous fossa, the posterior scapular and the suprascapular arteries anastomose. In the infraspinous fossa, the dorsalis scapulæ anastomoses with the suprascapular artery. In the subscapular fossa, the subscapular artery is distributed.

The **anterior circumflex artery** passes beneath the coraco-brachialis muscle and the short head of the biceps, across the anterior surface of the surgical neck of the humerus, for the supply of the deltoid muscle. As the vessel passes across the bicipital groove, it gives off the **bicipital artery,** which passes upward in that groove to supply the head of the humerus and the shoulder joint.

The **posterior circumflex artery** passes through the quadrangular space at the back of the shoulder (see p. 131), in company with the circumflex nerve. It passes around the posterior aspect of the surgical neck of the humerus and anastomoses with the anterior circumflex artery.

The **crucial anastomosis of the shoulder** is formed by the anterior circumflex artery, the posterior circumflex artery, the ascending branch of the superior profunda artery, and the acromial branch of the acromio-thoracic artery. (Morris, p. 542; Gray, p. 589.)

THE BRACHIAL ARTERY.

At the lower border of the tendon of the teres major muscle, the axillary artery becomes the brachial artery. The **brachial artery** passes down the arm, in a course represented by the lower portion of a line drawn from the middle of the clavicle to the middle of the bend of the elbow, when the arm is at right angles to the body. About one-half inch below the bend of the elbow, it divides into its terminal branches. In its course it is overlapped by the inner border of the biceps muscle.

RELATIONS.—The ulnar nerve, the internal cutaneous nerve, the basilic vein, and one of the venæ comites lie to its inner side, above. The biceps and the coraco-brachialis muscles and one of the venæ comites lie to its outer side. In front it is covered by the skin and the fascias of the arm and, slightly, by the biceps muscle. Behind, it is in relation with the triceps muscle, the musculo-spiral nerve, and the superior profunda artery. The median nerve lies, first, to its outer side, then above, and finally, to its inner side. At the bend of the elbow, the brachial artery passes through the cubital fossa. Here, the tendon of the biceps lies to its outer side and the median nerve to its inner side. It rests on the brachialis anticus muscle.

The **cubital fossa** is bounded, *externally*, by the supinator longus muscle; *internally*, by the pronator radii teres muscle; and *above*, by a line drawn between the two condyles of the humerus. *The floor* is formed by the brachialis anticus and the supinator brevis muscles. It contains the *musculo-spiral nerve*, the *tendon of the biceps muscle*, the *brachial artery*, and the *median nerve*, in the order given, from without inward. In the superficial fascia which covers in this space, the median basilic vein is found. It is separated from the artery by the bicipital fascia.

The **branches** of the brachial artery are: (1) the *superior profunda,* (2) the *inferior profunda,* (3) the *muscular,* (4) the *nutrient,* (5) the *anastomotica magna,* (6) the *radial,* and (7) the *ulnar.*

The **superior profunda artery** passes around the musculo-spiral groove, in company with the musculo-spiral nerve. It sends an ascending branch to the crucial anastomosis and terminates in the anastomosis about the elbow joint, after piercing the external intermuscular septum.

The **inferior profunda artery** passes, in company with the ulnar nerve, through the internal intermuscular septum and helps to form the elbow anastomosis.

The **nutrient artery** passes through a foramen in the shaft of the humerus and is distributed to that bone.

The **anastomotica magna artery** passes inward and helps to form the anastomosis around the elbow.

The **radial artery** is a branch of the brachial artery. It is given off in the cubital fossa, about one-half inch below the bend of the elbow. It passes outward and downward resting successively on the following muscles: (1) the tendon of the biceps, (2) the supinator brevis, (3) the flexor sublimis digitorum, (4) the pronator radii teres, (5) the flexor longus pollicis, and (6) the pronator quadratus. It passes beneath the extensor tendons of the thumb, to the dorsal aspect of the hand. It passes between the two heads of the first dorsal interosseous muscle, through the first interosseous space, into the palm of the hand, where it joins with the deep branch of the ulnar artery, to form the deep palmar arch. The vessel, in the forearm, is overlapped by the supinator longus muscle and, in the middle third of the forearm has the radial nerve to its outer side.

The **branches** of the radial artery are: (1) the *radial recurrent*, (2) the *muscular*, (3) the *anterior carpal*, (4) the *superficialis volæ*, (5) the *posterior carpal*, (6) the *metacarpal*, (7) the *dorsalis pollicis*, (8) the *dorsalis indicis*, (9) the *radialis indicis*, and (10) the *princeps pollicis*.

The **radial recurrent artery** passes backward, between the supinator longus and the brachialis anticus muscles, to help form the anastomosis around the elbow joint.

The **anterior carpal artery** passes inward, resting on the anterior surfaces of the carpal bones, to form the anterior carpal arch, by anastomosing with the anterior carpal branch of the ulnar artery.

The **superficialis volæ artery** passes over the anterior annular ligament of the wrist and over the muscles forming the thenar eminence, to anastomose with the superficial branch of the ulnar artery, completing the superficial palmar arch. This vessel is frequently absent.

The **posterior carpal artery** passes across the posterior aspects of the carpal bones and anastomoses with the posterior carpal branch of the ulnar artery to form the posterior carpal arch.

The **metacarpal artery** passes downward in the second interosseous space on the dorsum of the hand and, at the web of the fingers, divides into dorsal digital branches, which are distributed to the adjacent sides of the second and third fingers.

The **dorsalis pollicis artery** supplies the dorsal aspect of the thumb.

The **dorsalis indicis artery** supplies the radial side of the dorsal aspect of the index finger.

The **radialis indicis artery** is distributed to the radial side of the palmar surface of the index finger.

The **princeps pollicis artery** supplies the palmar aspect of the thumb.

The **ulnar artery** is one of the terminal branches of the brachial artery. It is given off in the cubital fossa. It passes inward and downward, lying beneath the pronator radii teres and the flexor sublimis digitorum muscles. When it reaches the inner portion of the forearm it passes directly downward, resting on the flexor profundus digitorum muscle and covered by the flexor carpi ulnaris muscle. In this part of its course the ulnar nerve lies to its inner side. It passes over the anterior annular ligament, between the pisiform bone and the unciform process of the unciform bone, into the palm of the hand, where it breaks up into its terminal branches.

The **branches** of the ulnar artery are: (1) the *anterior ulnar recurrent*, (2) the *posterior ulnar recurrent*, (3) the *com-*

mon interosseous, (4) the *muscular,* (5) the *anterior carpal,* (6) the *posterior carpal,* (7) the *superficial,* and (8) the *deep.*

The **anterior ulnar recurrent artery** passes backward, between the pronator radii teres and the brachialis anticus muscles, and enters into the formation of the elbow anastomosis.

The **posterior ulnar recurrent artery** passes backward, between the flexor sublimis digitorum and the flexor profundus digitorum muscles, to the anastomosis about the elbow joint.

The **common interosseous artery** is a branch of the ulnar artery as that vessel passes beneath the pronator radii teres muscle. It divides into the *anterior interosseous artery* and the *posterior interosseous artery.*

The **anterior interosseous artery** passes downward, lying on the anterior surface of the interosseous membrane, in company with the anterior interosseous nerve. It gives off branches to the muscles in its course; a branch which accompanies the median nerve, the **median artery;** the nutrient vessels to the radius and the ulna; and, at the upper border of the pronator quadratus muscle, divides into an *anterior branch* and a *posterior branch.* The **anterior branch** joins the anterior carpal arch. The **posterior branch** pierces the interosseous membrane and passes downward to join the posterior carpal arch.

The **posterior interosseous artery** passes between the radius and the ulna, above the interosseous membrane, and below the oblique ligament, to the posterior aspect of the forearm. Here it lies, in company with the posterior interosseous nerve, between the superficial and the deep layers of extensor muscles. It supplies the muscles in its course and ends by anastomosing with the posterior branch of the anterior interosseous artery. In its course it gives off the **interosseous recurrent artery,** which passes upward, between the supinator brevis and the anconeus muscles, to the anastomosis around the elbow.

The **anterior carpal artery** rests on the anterior surfaces of the carpal bones and joins with the anterior carpal
branch of the radial artery to form the anterior carpal
arch.

The **posterior carpal artery** passes across the posterior
aspect of the carpus and joins with the posterior carpal
branch of the radial to form the posterior carpal arch.

The **superficial branch** of the ulnar artery passes across
the hand, just beneath the palmar fascia, to form the superficial palmar arch by anastomosing with the superficialis volæ
branch of the radial artery.

The **deep** or **communicating branch** of the ulnar
artery passes deeply into the palm of the hand, between the
abductor minimi digiti and the flexor brevis minimi digiti
muscles, to join with the radial artery to complete the deep
palmar arch.

The **anastomosis about the elbow joint** is formed as
follows: in front of the external condyle, the anterior branch
of the superior profunda joins with the anterior branch of
the radial recurrent. Behind the external condyle, the posterior
branches of the radial recurrent and of the superior profunda
unite with a branch from the interosseous recurrent. In front
of the internal condyle, the anterior branch of the anastomotica
magna, the anterior branch of the inferior profunda, and the
anterior ulnar recurrent join. Behind the internal condyle, the
posterior ulnar recurrent anastomoses with the posterior
branch of the anastomotica magna artery and the posterior
branch of the inferior profunda artery. Behind the olecranon,
the interosseous recurrent, the anastomotica magna, the posterior ulnar recurrent, and the posterior branch of the inferior
profunda form an anastomotic arch.

The **superficial palmar arch** is formed by the anastomosis of the superficial branch of the ulnar artery and the
superficialis volæ branch of the radial artery. It passes transversely across the palm of the hand, in a line drawn from
the ball of the thumb, when the thumb is held at right

angles to the hand. It rests upon the branches of the median nerve, which separate it from the tendons of the flexor sublimis digitorum muscle. It is covered by the palmar fascia. The **branches** of the superficial palmar arch are the *four digital arteries.* The **first digital artery** passes to the inner side of the little finger. The **second, third, and fourth digital arteries** pass downward in the line of the interosseous spaces and, at the web of the fingers, divide into **collateral digital branches** which supply the adjacent sides of the fifth and fourth, the fourth and third, and the third and second fingers, respectively. The outer side of the second finger is supplied by the radialis indicis artery. The thumb is supplied by the princeps pollicis artery.

The **deep palmar arch** is formed by the radial artery and the deep branch of the ulnar artery. It passes transversely across the palm of the hand about one-half inch above the superficial palmar arch. It rests upon the anterior surfaces of the metacarpal bones and on the palmar interosseous muscles. The **branches** of the deep palmar arch are: (1) the *palmar interossei,* (2) the *perforating,* and (3) the *palmar recurrent.*

The **palmar interossei arteries,** three in number, pass forward between the metacarpal bones and, in the web of the fingers, pass toward the palmar surface to anastomose with the second, third, and fourth digital branches of the superficial palmar arch, just before those vessels divide into their collateral digital branches.

The **perforating arteries,** three in number, pass toward the dorsum of the hand, between the metacarpal bones and between the two heads of the second, third, and fourth dorsal interosseous muscles, to anastomose with the dorsal interosseous branches of the posterior carpal arch and with the metacarpal branch of the radial artery.

The **palmar recurrent arteries,** three in number, pass backward to the anterior carpal arch.

The **posterior carpal arch** is formed by the posterior

carpal branch of the radial artery and the posterior carpal branch of the ulnar artery. From this arch **two dorsal interosseous arteries** are given off, which pass in the third and the fourth interosseous spaces, to divide, at the web of the fingers, into the **dorsal digital arteries** for the supply of the dorsal aspects of the adjacent sides of the fifth and fourth, and the fourth and third fingers. The adjacent sides of the third and second fingers are supplied by the dorsal digital branches of the metacarpal branch of the radial artery. The outer side of the second finger is supplied by the dorsalis indicis artery. The dorsal aspect of the thumb is supplied by the dorsalis pollicis artery. The inner side of the little finger is supplied by branches which come around from the first digital artery. The collateral digital arteries send twigs to the dorsal surfaces of the second, third, fourth, and fifth fingers. (Morris, p. 548; Gray, p. 593.)

THE RELATION OF THE STRUCTURES IN THE PALM OF THE HAND.

Between the skin and the metacarpal bones, the structures in the palm of the hand are placed in the following order: (1) the skin, (2) the superficial fascia, (3) the palmar fascia, (4) the superficial palmar arch, (5) the median nerve, (6) the tendons of the flexor sublimis digitorum muscle, (7) the tendons of the flexor profundus digitorum muscle and the lumbrical muscles, (8) the deep palmar arch, (9) the palmar interosseous muscles, and (10) the metacarpal bones.

CHAPTER IX.

THE ABDOMINAL PARIETIES.

The superficial fascia of the abdominal wall is divisible into two layers; the superficial layer of the superficial fascia or the fascia of Camper, and the deep layer of the superficial fascia or fascia of Scarpa. The superficial layer of the superficial fascia of the abdomen is well supplied with fat. Between the superficial and the deep layers we find the terminations of the lower intercostal and of the lumbar arteries, twigs from the deep epigastric and the superior epigastric arteries, and the superficial epigastric and the superficial circumflex iliac arteries, from the common femoral artery. The nerves in this region are derived from the lower intercostal nerves and from the ilio-hypogastric branch of the lumbar plexus. The superficial layer of the superficial fascia is continuous with the superficial fascia of the thigh. The deep layer of the superficial fascia is attached to the fascia lata, just below Poupart's ligament, to the symphisis pubis, and to the spine of the pubes. Between the two latter points of attachment the deep layer of the superficial fascia is prolonged downward into the scrotum and becomes continuous with the dartos. This arrangement forms a passage from the abdomen to the scrotum, which is termed the scroto-abdominal passageway. Through this opening, collections of fluid may pass from the tissues of the scrotum to the abdomen. (Morris, p. 421; Gray, p. 447.)

The deep fascia of the abdomen is intimately attached to the aponeurosis of the external oblique muscle.

The abdominal walls derive their strength largely from the muscles and the aponeuroses which form them. These muscles are: (1) the *external oblique*, (2) the *internal oblique*, (3) the *transversalis*, (4) the *rectus*, and (5) the *pyramidalis*.

148

The **external oblique muscle** is inserted by a broad aponeurosis which is termed the **aponeurosis of the external oblique muscle.** This aponeurosis meets with the aponeurosis of the corresponding muscle of the opposite side of the body to form a thickened ridge, which extends from the ensiform cartilage of the sternum to the symphisis pubis. This raphé is spoken of as the **linea alba.** From the anterior superior spine of the ilium to the spine of the pubes, the aponeurosis of the external oblique muscle extends in a thickened band which is known as **Poupart's ligament.** From the spine of the pubes, the tissue forming Poupart's ligament is reflected along the ilio-pectineal line, forming a triangular band of fibrous tissue which is known as **Gimbernat's ligament.** From the spine of the pubes, a triangular process of the external oblique aponeurosis, termed the **triangular ligament of the abdomen,** passes upward and inward to the linea alba.

The **external abdominal ring** is a triangular separation in the fibres of the aponeurosis of the external oblique muscle, which permits the passage of the spermatic cord into the scrotum. The external abdominal ring is bounded, *externally*, by the external pillar; *internally*, by the internal pillar; and *below,* by the crest of the pubes. The **external pillar** corresponds to the inner portion of Poupart's ligament and is attached to the spine of the pubes. The **internal pillar** is inserted into the symphisis pubis. Running between the external and the internal pillars of the external abdominal ring, delicate fibres may be observed which are spoken of as the **intercolumnar fibres.** The **intercolumnar fascia** is a connective tissue membrane which takes its origin from the pillars of the external abdominal ring and which passes downward to cover the spermatic cord. (Morris, pp. 424 and 1135; Gray, pp. 448 and 1181.)

The **aponeurosis of the internal oblique muscle** passes inward until it comes to the outer border of the rectus muscle. In the upper three-fourths of its extent, it then divides into two layers, one of which passes behind, and the other of which passes in front of the rectus muscle. In the lower one-fourth of its extent, the aponeurosis of the internal oblique muscle does not divide; but passes in a single layer in front

of the rectus muscle. The position of division of the internal
oblique aponeurosis is indicated by a curved line which extends, with its convexity outward, from the costal cartilage of
the eighth rib to the spine of the pubes. This line is known
as the **linea semilunaris.** The fibres of the internal oblique
muscle which arise from Poupart's ligament, arch over the
inguinal canal to be inserted, in common with similar fibres
from the transversalis muscle, into the spine of the pubes by
a tendinous structure which is known as the **conjoined tendon** of the internal oblique and transversalis muscles. The
fibres of the internal oblique muscle which arch over the
inguinal canal are known as the **arching fibres** of the internal
oblique.

As the internal oblique muscle passes over the inguinal
canal, it gives off fibres which pass downward, covering the
spermatic cord. These fibres constitute the **cremaster muscle.**
(Morris, p. 427; Gray, pp. 451 and 1184.)

The **aponeurosis of the transversalis** muscle, in the
upper three-fourths of its extent, passes behind the rectus muscle
to be inserted into the linea alba. In the lower one-fourth
of its course, the transversalis aponeurosis passes in front of
the rectus muscle. The fibres of the transversalis muscle
which arise from Poupart's ligament arch over the inguinal
canal, **arching fibres** of the transversalis, and are inserted by
the **conjoined tendon** into the spine of the pubes, in common with the arching fibres of the internal oblique muscle.
(Morris, p. 429; Gray, pp. 453 and 1184.)

The **rectus muscle** is enclosed in a **sheath** which, in
its upper three-fourths, is formed, anteriorly, by the aponeurosis of the external oblique and one-half the aponeurosis of
the internal oblique; posteriorly, by one-half the aponeurosis of
the internal oblique and the aponeurosis of the transversalis.
In the lower one-fourth, the anterior layer of the sheath of
the rectus muscle is formed by the aponeurosis of the external oblique, the internal oblique, and the transversalis muscles;
posteriorly, the rectus muscle rests upon the transversalis fascia.
The **semilunar fold of Douglas** is a thickened, crescentic
band which marks the point at which the aponeuroses of

the abdominal muscles pass in front of the rectus muscle. The deep epigastric artery enters the sheath of the rectus muscle at this point. After opening the sheath of the rectus muscle, two or three fibrous intersections are to be seen passing across the muscular substance; these are known as the **lineæ transversæ.** (Morris, p. 423; Gray, p. 453.)

The **transversalis fascia** is a broad sheet of connective tissue which separates the transversalis muscle from the preperitoneal fat. This fascia is a process of the lumbar fascia. The **lumbar fascia** is composed of three layers; the posterior layer arises from the tips of the spinous processes of the lumbar vertebræ; the middle layer arises from the tips of the transverse processes of the lumbar vertebræ; and the anterior layer arises from the anterior surfaces of the transverse processes of the lumbar vertebræ. The posterior and the middle layers include the extensor dorsi communis muscle between them; while the quadratus lumborum muscle is found between the middle and the anterior layers. At the anterior border of the quadratus lumborum muscle, these three layers become more or less completely fused. The posterior portion is then termed the **lumbar aponeurosis** and gives origin to the latissimus dorsi and to the abdominal muscles. The anterior portion is prolonged inward and forward as the transversalis fascia. The transversalis fascia is prolonged downward into the thigh, forming the anterior layer of the sheath of the femoral vessels. (Morris, p. 431; Gray, pp. 456 and 1185.)

The **internal abdominal ring** is an opening in the transversalis fascia which permits of the passage of the structures which form the spermatic cord. It is situated about one-half inch above the middle of Poupart's ligament. The deep epigastric artery lies just internal to its inner margin. The transversalis fascia is prolonged downward, through the internal abdominal ring and into the inguinal canal, as a funnel-shaped membrane which encloses the spermatic cord. It is termed the **infundibuliform fascia.** (Morris, pp. 432 and 1137; Gray, 456 and 1186.)

The transversalis fascia is separated from the peritoneum by a layer of areolar tissue containing fat, which is termed the **preperitoneal fat.**

The **peritoneum** is the serous membrane which lines the abdominal cavity. If the portion of this membrane which lines the anterior abdominal wall is examined, the urachus will be seen passing beneath it from the summit of the bladder to the umbilicus. External to the ridge formed by the underlying urachus, the ridge formed by the passage of the obliterated hypogastric artery may be seen, extending from the lateral aspect of the bladder to the umbilicus. Still more external, the ridge produced by the underlying deep epigastric artery, as it passes upward to enter the sheath of the rectus muscle, may be seen. Between the ridges produced by the underlying structures we have the three inguinal fossæ. The **internal inguinal fossa** lies between the urachus and the obliterated hypogastric artery; the **middle inguinal fossa** lies between the obliterated hypogastric artery and the deep epigastric artery; and the **external inguinal fossa** lies outside the deep epigastric artery. The middle inguinal fossa is unequally divided into an outer and an inner portion by the outer margin of the rectus muscle. The outer and larger portion of this space, bounded, *externally*, by the deep epigastric artery; *internally*, by the outer margin of the rectus muscle; and *below*, by Poupart's ligament, is called **Hesselbach's triangle.** A direct hernia passes through this space. The external inguinal fossa presents a dimple which marks the position of the internal abdominal ring. An indirect hernia passes through this space. (Morris, p. 1139; Gray, pp. 1186 and 1190.)

The **inguinal canal** is an oblique passageway through the anterior abdominal wall, which begins at the internal abdominal ring and ends at the external abdominal ring. The anterior wall is formed, for the outer two-thirds of its extent, by the skin, the superficial fascia, the aponeurosis of the external oblique, the internal oblique muscle, and the transversalis muscle. The anterior wall for the inner one-third of the extent of the canal, is formed by the skin, the superficial fascia, and the aponeurosis of the external oblique muscle. The posterior wall for the outer two-thirds of the extent of the canal, is formed by the peritoneum, the preperitoneal fat, and the transversalis fascia. The posterior wall of the inner one-third of the extent of the canal, is formed by the peritoneum, the preperitoneal fat, the transversalis fascia, the conjoined tendon of the internal oblique and transversalis muscles, and a portion of the triangular ligament of the abdomen. The arching fibres of the internal oblique and the transversalis muscles pass over the superior border of the canal. The transversalis fascia is reflected into the inner opening of the canal, as the infundibuliform fascia. The intercolumnar fascia covers over the external opening of the canal. The inguinal canal contains the spermatic cord, in the male, and the round ligament in the female. (Morris, p. 1137; Gray, p. 1185.)

If a hernia enters the internal abdominal ring and passes through the inguinal canal, leaving it by passing through the external abdominal ring, it is called an **external, indirect,** or **oblique inguinal hernia.** Such a hernia lies outside the deep epigastric artery. It would be covered by the following structures, passing from without inward: (1) the skin, (2) the superficial fascia, (3) the intercolumnar fascia, (4) the cremaster muscle, (5) the infundibuliform fascia, (6) the preperitoneal fat, and (7) the peritoneum. In cutting the constriction of such a hernia, the incision should pass upward and outward to avoid wounding the deep epigastric artery.

If a hernia passes through Hesselbach's triangle, across the abdominal wall, and out from the external abdominal ring, it is called a **direct** or an **internal inguinal hernia.** Such a hernia lies internal to the deep epigastric artery. It would be covered by the following structures, passing from without inward: (1) the skin, (2) the superficial fascia, (3) the intercolumnar fascia, (4) the conjoined tendon of the internal oblique and the transversalis muscles, (5) the transversalis fascia, (6) the preperitoneal fat, and (7) the peritoneum. In case the triangular ligament of the abdomen were well developed, it would form one of the coverings of such a hernia and would be met with between the intercolumnar fascia and the conjoined tendon. In cutting the constriction of such a hernia, the incision should pass upward and inward to avoid wounding the deep epigastric artery. (Morris, p. 1138; Gray, p. 1186.)

CHAPTER X.

THE LOWER EXTREMITY.

The **superficial fascia** of the thigh is continuous with that of the abdomen, of the perineum, and of the leg. In the upper part of the thigh, the superficial fascia consists of two layers. The deep layer of the superficial fascia of the thigh, which covers the saphenous opening in the fascia lata, is termed the **cribriform fascia.** It is attached to the margins of the saphenous opening.

In the superficial fascia covering the dorsum of the foot, we find a venous arch, which terminates at either end in a large vein. At the inner extremity of this arch, the **internal** or **long saphenous vein** begins. This vein passes in front of the internal malleolus and ascends in the superficial fascia of the leg, in company with the long saphenous nerve. At the knee, the long saphenous vein leaves the nerve and passes behind the internal condyle of the femur; it then ascends in the superficial fascia of the thigh to empty, finally, into the femoral vein, about an inch below Poupart's ligament. In order to reach the femoral, the long saphenous vein passes through the saphenous opening in the fascia lata.

The **external** or **short saphenous vein** begins at the outer termination of the venous arch on the dorsum of the foot. It passes behind the external malleolus and up the posterior aspect of the leg, in company with the short saphenous nerve. It pierces the deep fascia which covers in the popliteal space and empties into the popliteal vein. (Morris, pp. 363 and 669; Gray, pp. 506 and 670.)

The **deep fascia** of the thigh is known as the **fascia lata.** It is composed of a portion which is attached to the ilium and which passes inward toward the pubes, known as the **iliac portion** of the fascia lata, and a portion which is attached to the pubes and which passes outward toward the iliac portion, known as the **pubic portion** of the fascia lata. The pubic portion of the fascia lata occupies a plane somewhat posterior to that occupied by the iliac portion. At the position of junction of the two processes, in the upper part of Scarpa's triangle, there is a cleft which is known as the saphenous opening. The long saphenous vein passes through this opening to reach the femoral vein. The curved, superior margin of the saphenous opening is known as **Hay's ligament** or the **falciform process of the fascia lata.**

The **ilio-tibial band** is a thickened portion of the fascia lata, which extends from the crest of the ilium to the outer tuberosity of the tibia. It receives the insertion of the tensor vaginæ femoris muscle. The gluteus maximus muscle is also inserted largely into the fascia lata. (Morris, pp. 364 and 1142; Gray, pp. 506 and 1193.)

The **deep fascia of the leg** is attached to the anterior and the internal borders of the tibia, so that there is no deep fascia covering the internal surface of the shaft of that bone, the bone being entirely subcutaneous.

The **popliteal fascia** is the fascia covering the popliteus muscle.

The **deep transverse fascia** of the leg is a process of the deep fascia of that region, which separates the soleus muscle from the tibialis posticus, the flexor longus hallucis, and the flexor longus digitorum muscles. (Morris, p. 388; Gray, p. 520.)

The **interosseous membrane** is a dense band of fibrous tissue, which stretches between the interosseous border of the fibula and the interosseous border of the tibia.

The **anterior annular ligament** of the ankle is a thickening of the deep fascia of the leg, which is attached to the tibia, internally, and to the fibula, externally. The tendons of the tibialis anticus, the extensor longus digitorum, and the extensor proprius hallucis muscles, the anterior tibial artery, and the anterior tibial nerve pass beneath it to enter the foot.

The **internal annular ligament** of the ankle is a thickening of the deep fascia of the leg, which is attached, above,

to the internal malleolus; and below, to the inner margin
of the os calcis. Beneath it, the tendons, vessels, and nerves
pass to reach the sole of the foot in the following order
from above downward: (1) the tendon of the tibialis posticus
muscle, (2) the tendon of the flexor longus digitorum muscle,
(3) the posterior tibial artery with one of the posterior
tibial veins on either side, (4) the posterior tibial nerve,
and (5) the tendon of the flexor longus hallucis muscle.

The **external annular ligament** of the ankle is attached,
above, to the external malleolus; and below, to the outer border
of the os calcis. The tendons of the peroneus longus and the
peroneus brevis muscles pass beneath it, in order to reach the
sole of the foot. (Morris, p. 388; Gray, p. 528.)

The deep fascia of the sole of the foot is known as the
plantar fascia. It is attached, posteriorly, to the inner tubercle
of the os calcis and passes forward toward the web of the toes.
The central portion of the plantar fascia is of great thickness,
while the lateral portions are thinner. Opposite the metatarso-
phalangeal articulations, the central portion of the plantar fascia
divides into five processes which pass forward in the line of
the metatarsal bones. Between these processes the digital
vessels and nerves pass to the toes. These five processes
send off slips, which are attached to the skin on the sole of
the foot, and then divide into two secondary processes which
allow the flexor tendons to pass forward to the toes. These
secondary processes are intimately connected with the sheaths
of the flexor tendons. (Morris, p. 397; Gray, p. 529.)

The space between Poupart's ligament and the ilio-pectineal
line is known as the **crural arch.** A process of fibrous tissue
which extends from Poupart's ligament to the ilio-pectineal line
is known as the **ilio-pectineal ligament.** This ligament
divides the crural arch into an outer, muscular compartment,
and an inner, vascular compartment. The iliacus and the psoas
magnus muscles and the external cutaneous and the anterior
crural nerves pass through the muscular compartment. The
vascular compartment is occupied by the femoral artery, the
femoral vein, and the femoral canal. In the vascular compart-

ment the femoral vessels are contained in a sheath, the **sheath of the femoral vessels,** which is formed, in front, by the transversalis fascia; and behind, by the iliac fascia. This sheath is divided by septa of connective tissue into an outer compartment, which contains the femoral artery, a middle compartment, which contains the femoral vein, and an inner compartment, which is normally empty, except for a few lymphatics. The innermost compartment of the femoral sheath is known as the **femoral canal.** It is about one-half inch in length, extending from the femoral ring to the saphenous opening.

The **femoral ring** is the opening into the femoral canal beneath the crural arch. It is bounded, *above*, by Poupart's ligament; *below*, by the ilio-pectineal line; *internally*, by Gimbernat's ligament; and *externally*, by the femoral vein. It is closed by a pad of fat and a lymphatic gland. This tissue is a process of the preperitoneal fat and is known as the **septum crurale.**

The **saphenous opening** and its coverings have been described on page 155.

If a hernia passes through the femoral ring, the femoral canal, and the saphenous opening, it is known as a **femoral hernia.** Such a hernia would be covered by (1) the skin, (2) the superficial fascia, (3) the cribriform fascia, (4) the femoral sheath, (5) the septum crurale, (6) the preperitoneal fat, and (7) the peritoneum. The deep epigastric artery passes along the superior and external borders of the femoral ring. At times an obturator artery, arising anomalously from the external iliac artery or the deep epigastric artery, passes across the femoral ring and would be liable to injury in the operation for the relief of strangulation of this kind of hernia. (Morris, p. 1141; Gray, p. 1191.)

SCARPA'S TRIANGLE.

Scarpa's triangle is a triangular space situated in the upper portion of the anterior aspect of the thigh. It is bounded, *above*, by Poupart's ligament; *internally*, by the adductor longus muscle; and *externally*, by the sartorius muscle.

The floor is formed by the iliacus, the psoas magnus, the pectineus, and the adductor brevis muscles. It contains the *anterior crural nerve*, and its branches, the *femoral artery* and its branches, and the *femoral vein* and its tributaries, in the order given from without inward. (Morris, p. 1189; Gray, p. 630.)

Hunter's canal is a musculo-membranous space which is situated between the adductor magnus and the vastus internus muscles. The roof of this canal lies in a vertical plane and is formed by an aponeurotic slip which passes between the two muscles. It contains the *femoral artery*, the *femoral vein*, the *anastomotica magna artery*, and the *long saphenous nerve*. (Morris, p. 1190; Gray, p. 630.)

THE LUMBAR PLEXUS.

The **lumbar plexus** is formed by the anterior divisions of the first four lumbar nerves. It is formed in the substance of the psoas magnus muscle; its branches making their appearance through the fibres of that muscle.

The **branches** of the lumbar plexus are: (1) the *ilio-hypogastric*, (2) the *ilio-inguinal*, (3) the *external cutaneous*, (4) the *genito-crural*, (5) the *anterior crural,* (6) the *obturator*, and (7) the *accessory obturator*.

The **ilio-hypogastric nerve** is a branch of the first lumbar nerve. It makes its appearance beneath the outer border of the psoas magnus muscle, passes across the quadratus lumborum muscle, and, at the crest of the ilium, pierces the transversalis muscle. It then runs forward, between the transversalis and the internal oblique muscles. About two inches behind the anterior superior spine of the ilium the nerve gives off an *iliac branch* and a *hypogastric branch*. The **iliac branch** pierces the internal oblique and the external oblique muscles and passes over the crest of the ilium to supply the skin of the gluteal region. The **hypogastric branch** pierces the internal oblique muscle and runs forward, between it and the external oblique muscle, to a point about an inch above the external abdominal ring. Here it pierces the aponeurosis of the external oblique muscle and is distributed to the skin of the hypogastric region.

The **ilio-inguinal nerve** is a branch of the first lumbar nerve. It makes its appearance at the outer border of the psoas magnus muscle. It then passes across the quadratus lumborum and the iliacus muscles, to pierce the transversalis muscle at the crest of the ilium. It then passes forward, between the internal oblique and the transversalis muscles. About opposite the anterior superior spinous process of the ilium it pierces the internal oblique muscle and continues in its forward course between it and the aponeurosis of the external oblique muscle. It passes through the external abdominal ring, lying superficial to the spermatic cord, and is distributed to the skin on the adductor surface of the thigh and to the scrotum, in the male, or the labium major, in the female.

The **external cutaneous nerve** is a branch of the second and third lumbar nerves. It makes its appearance at the outer border of the psoas magnus muscle and passes across the iliacus muscle to leave the pelvis by passing above the tendon of origin of the sartorius muscle, just below the anterior superior spinous process of the ilium. It then divides into an *anterior branch* and a *posterior branch*. The **anterior branch** pierces the fascia lata at about the junction of the middle and lower thirds of the thigh and is distributed to the skin on the outer aspect of the thigh as far down as the knee. The **posterior branch** supplies the skin of the outer portion of the posterior aspect of the thigh.

The **genito-crural nerve** is a branch of the first and second lumbar nerves. The nerve makes its appearance on the superior surface of the psoas magnus muscle and, in relation with the external iliac artery, divides into a *genital branch* and a *crural branch*. The **genital branch** passes into the inguinal canal, through the internal abdominal ring and forms one of the constituents of the spermatic cord. It is distributed to the cremaster muscle in the male. In the female it is found passing with the round ligament of the uterus. The **crural branch** passes along the superior surface of the external iliac artery and enters the thigh by passing beneath Poupart's ligament. It is distributed to the skin covering the upper portion of Scarpa's triangle.

The **anterior crural nerve** is a branch of the second, third, and fourth lumbar nerves. It is found lying parallel with and beneath the outer border of the psoas magnus muscle. It passes beneath Poupart's ligament, through the muscular compartment of the crural arch, between the iliacus and the psoas magnus muscles, into the thigh. In Scarpa's triangle the anterior crural nerve lies to the outer side of the femoral artery. The **branches** of the anterior crural nerve are: (1) the *middle cutaneous*, (2) the *internal cutaneous*, (3) the *long saphenous*, (4) the *muscular*, and (5) the *articular*.

The **middle cutaneous nerve** is a branch of the anterior crural nerve in Scarpa's triangle. It pierces the sartorius muscle about four inches below Poupart's ligament and is distributed to the skin of the anterior portion of the thigh by an external and an internal branch.

The **internal cutaneous nerve** is a branch of the anterior crural nerve in Scarpa's triangle. It passes across the femoral artery and pierces the fascia lata at the junction of the middle and lower thirds of the thigh. It is distributed to the skin on the inner aspect of the thigh by an anterior and a posterior branch.

The **long saphenous nerve** is a branch of the anterior crural nerve in Scarpa's triangle. It lies to the outer side of the femoral artery as they pass together through Scarpa's triangle and Hunter's canal. It leaves Hunter's canal in company with the anastomotica magna artery, by piercing the roof of that space. It is then found between the sartorius muscle and the gracilis muscle. It passes behind the internal condyle of the femur and enters the leg. It passes in the superficial fascia of the leg, in company with the long saphenous vein, to terminate in the skin over the ball of the great toe. In its course, the long saphenous nerve gives the **nervus cutaneous patellæ** to the prepatellar plexus. This branch pierces the tendon of the sartorius muscle. It also gives branches to the skin of the inner side of the leg.

The **muscular branches** of the anterior crural nerve supply the iliacus, the pectineus, the sartorius, the rectus femoris, the vastus externus, the vastus internus, and the crureus muscles.

The **articular branches** are distributed to the hip joint and to the knee joint. They are usually given off from some of the muscular branches. That to the hip from the nerve to the rectus femoris; that to the knee joint from the nerve to the crureus.

The **obturator nerve** is a branch of the third and fourth lumbar nerves. It passes along the inner border of the psoas magnus muscle, lying between it and the external iliac vein, to the obturator foramen. It passes through the obturator foramen, in company with the obturator vessels, into the thigh and divides into an *anterior branch* and a *posterior branch*. The **anterior branch** lies between the adductor brevis and the pectineus muscles. It gives off a branch to the hip joint, a cutaneous branch, and supplies the adductor longus, the adductor brevis, and the gracilis muscles. The **posterior branch** lies between the adductor brevis and the adductor magnus muscles. It supplies the obturator externus and the adductor magnus muscles and gives branches to the hip and to the knee joints.

The **accessory obturator nerve** is frequently wanting. When it is present it passes through the obturator foramen and is distributed to the hip joint. (Morris, p. 826; Gray, p. 850.)

The **subsartorial plexus** lies beneath the inner border of the sartorius muscle. It is formed by branches from the internal cutaneous nerve, the long saphenous nerve, and the obturator nerve.

The **prepatellar plexus** is found in the subcutaneous tissue in front of the patella. It is formed by the external cutaneous, the internal cutaneous, the middle cutaneous, and a branch of the long saphenous nerves.

THE FEMORAL ARTERY.

The **femoral artery** may, for convenience of description, be divided into the *common femoral,* the *deep femoral,* and the *super-ficial femoral arteries.*

The **common femoral artery** is the continuation of the external iliac artery from beneath Poupart's ligament. It lies in Scarpa's triangle and, about two inches below Poupart's ligament, divides into the superficial femoral and the deep femoral. A line drawn from a point midway between the anterior superior spine of the ilium and the symphisis pubis to the adductor tubercle would represent the course of this vessel and of the superficial femoral artery.

RELATIONS.—The common femoral artery rests on the tendon of the psoas magnus muscle and on the pectineus muscle. The femoral vein lies to its inner side, the anterior crural nerve lies to its outer side, and it is covered by the skin and fascias of the thigh. It is contained in the femoral sheath, which is formed, in front, by the transversalis fascia, and behind, by the iliac fascia. The crural branch of the genito-crural nerve lies in front of this sheath.

The **branches** of the common femoral artery: (1) the *super-ficial epigastric,* (2) the *superficial circumflex iliac,* (3) the *super-ficial external pudic,* (4) the *deep femoral,* and (5) the *superficial femoral.*

The **superficial epigastric artery** becomes an occupant of the superficial fascia, just after it is given off from the common femoral artery. It passes over Poupart's ligament to supply the superficial fascia and the skin of the lower portion of the abdominal wall, nearly to the umbilicus.

The **superficial circumflex iliac artery** pierces the fascia lata and becomes an occupant of the superficial fascia of the thigh soon after it is given off. It passes parallel to Poupart's ligament and to the crest of the ilium, supplying the skin and superficial fascia of the abdomen in its course.

The **superficial external pudic artery** passes in the superficial fascia to the penis and the scrotum, in the male; to the labium major and the clitoris, in the female.

The **deep external pudic artery** passes across the pectineus and adductor longus muscles to the scrotum, in the male; to the labium major, in the female.

The **deep femoral artery** or **profunda femoris** is a branch of the common femoral in Scarpa's triangle, about two inches below Poupart's ligament. It passes inward and backward, beneath the adductor longus muscle, and lies close to the femur. It rests upon the iliacus, the pectineus, the adductor brevis, and the adductor magnus muscles. In the first part of its course, the femoral artery, the femoral vein, and the branches of the anterior crural nerve lie above it; but lower down, it is separated from these structures by the adductor longus muscle. The vastus internus muscle lies to its outer side.

The **branches** of the deep femoral artery are: (1) the *external circumflex*, (2) the *internal circumflex*, and (3) the *three perforating*.

The **external circumflex artery** passes outward beneath the sartorius and the rectus femoris muscles, resting upon the iliacus muscle. It divides into *ascending, descending* and *transverse branches*. The **ascending branch** is distributed to the muscles in the gluteal region and sends a branch to the hip. The **descending branch** passes downward to the anastomosis around the knee joint. It gives branches to the muscles in its course. The **transverse branch** passes around the neck of the femur and joins the crucial anastomosis of the thigh.

The **internal circumflex artery** passes between the psoas and pectineus muscles, then between the obturator externus and the adductor brevis muscles, and finally, between the quadratus femoris and the adductor magnus muscles to the posterior aspect of the femur. Here it joins with the transverse branch of the external circumflex artery, the superior perforating artery, and the sciatic artery to form the **crucial anastomosis** of the thigh. It gives branches to the muscles in its course and a branch to the hip joint.

The **perforating arteries,** three in number, are named the superior, the middle, and the inferior. These vessels pierce the adductor magnus muscle and supply the tissues on the posterior aspect of the thigh. The **superior perforating artery** passes through the adductor magnus muscle, just above the adductor brevis muscle. It helps to form the crucial anastomosis. The **middle perforating artery** passes through the adductor magnus muscle, after piercing the fibres of the adductor brevis muscle. The **inferior perforating artery** passes through the adductor magnus muscle, just below the adductor brevis muscle. It helps to form the anastomosis about the knee joint. These perforating arteries anastomose with each other, forming a series of arches. When the profunda femoris pierces the adductor magnus muscle, it is known as the **fourth perforating artery**.

The **superficial femoral artery** is given off from the common femoral artery, in Scarpa's triangle, about two inches below Poupart's ligament. It passes through Scarpa's triangle, lying above the deep femoral artery, and, at the apex of that triangle, enters Hunter's canal. It passes through an opening in the adductor magnus muscle and becomes an occupant of the popliteal space. It is then known as the popliteal artery.

RELATIONS.—The superficial femoral artery is separated from the deep femoral by the adductor longus muscle. The femoral vein, which was seen to the inner side of the common femoral artery, gradually passes behind the superficial femoral artery and, in Hunter's canal, lies to its outer side. The long saphenous nerve crosses above the superficial femoral artery and in Hunter's canal lies to its outer side. The internal cutaneous nerve crosses the superficial femoral artery from within outward.

The **branches** of the superficial femoral artery are: (1) the *muscular* and (2) the *anastomotica magna*.

The **muscular branches** supply the muscles in the course of the superficial femoral artery.

The **anastomotica magna artery** is given off in Hunter's canal and divides into a *superficial branch* and a *deep branch*. The **superficial branch** pierces the roof of this space, in company with the long saphenous nerve and goes to the anastomosis about the knee joint. The **deep branch** passes along the tendon of the adductor magnus muscle and goes to the anastomosis about the knee. (Morris, p. 602; Gray, p. 630.)

THE POPLITEAL SPACE.

The **popliteal space** is the name given to the region behind the knee joint. It is a lozenge-shaped space bounded, *externally and above*, by the biceps muscle (the outer hamstring); *externally and below*, by the outer head of the gastrocnemius muscle; *internally and above*, by the semimembranosus, and the semitendinosus muscles (the internal hamstring); and *internally and below*, by the inner head of the gastrocnemius muscle. *The floor* is formed by the posterior surface of the shaft of the femur, the posterior ligament of the knee joint, and the popliteus muscle. It is' covered by the skin, the superficial fascia, and the deep fascia. It contains the *external* and the *internal popliteal nerves*, the *popliteal vein*, and the *popliteal artery and its branches.*

THE POPLITEAL ARTERY.

The **popliteal artery** is the continuation of the femoral artery, after the latter has passed through the opening in the adductor magnus muscle. It passes through the popliteal space as the most posterior structure and, at the lower border of the popliteus muscle, divides into its terminal branches.

RELATIONS.—Above, the internal popliteal nerve and the popliteal vein are found, in the order named from above downward. The vein lies to the outer side of the artery in the upper portion of the popliteal space, but crosses above it and, in the lower portion of the popliteal space, lies to its inner side. The nerve is the most superficial structure and also crosses the artery from without inward and from above downward. Below, the artery rests upon the posterior surface of the shaft of the femur, the posterior ligament of the knee joint, and the popliteus muscle. These relations are given as the dissection is made, with the body in the prone position. In the upright posture, the structures which are enumerated as being above the vessel are found behind it, while those which are placed below are in front of it.

The **branches** of the popliteal artery are: (1) the *superior external articular*, (2) the *superior internal articular*, (3) the *inferior external articular*, (4) the *inferior internal articular*, (5) the *azygos articular* (6) the *muscular*, (7) the *cutaneous*, (8) the *anterior tibial*, and (9) the *posterior tibial*.

The **superior external articular artery** passes beneath

the biceps muscle and the outer head of the gastrocnemius muscle to the knee joint anastomosis.

The **superior internal articular artery** passes above the inner head of the gastrocnemius and beneath the semimembranosus muscle to the anastomosis about the knee joint.

The **inferior external articular artery** passes beneath the external head of the gastrocnemius and the tendon of the biceps and above the head of the fibula and the popliteus muscle to the anastomosis about the knee joint.

The **inferior internal articular artery** passes along the popliteus muscle to the knee joint anastomosis.

The **azygos articular artery** pierces the posterior ligament of the knee joint and supplies the joint.

The **muscular branches** supply the muscles which form the boundaries of the popliteal space.

The **cutaneous branches** are distributed to the skin covering the popliteal space and to the skin over the calf of the leg. One branch, larger than its fellows, usually follows the short saphenous nerve. (Morris, p. 609; Gray, p. 638.)

The **anastomosis around the knee joint** consists of three arches. The superior arch lies above the patella and is formed by the superior external articular, the descending branch of the external circumflex, the anastomotica magna, and the end of the deep femoral arteries. The middle arch lies just below the patella and is formed by the inferior external articular, the superior internal articular, and the anastomotica magna arteries. The inferior arch lies a short distance below the middle arch and is formed by the tibial recurrent and the inferior internal articular arteries. The inferior arch anastomoses with the middle arch.

The **anterior tibial artery** is a branch of the popliteal artery in the popliteal space. It is given off opposite the lower border of the popliteus muscle and passes between the tibia and the fibula, between the two heads of the tibialis posticus muscle, and above the interosseous membrane, to the anterior aspect of the leg. It passes down the leg, resting upon the interosseous membrane, first, between the tibialis anticus and the extensor longus digitorum muscles; second, between the tibialis anticus and the extensor proprius hallucis muscles; and third, between the extensor proprius hallucis and the extensor longus digitorum muscles. It then passes beneath the anterior annular ligament of the ankle to the dorsum of the foot, where it becomes the dorsalis pedis artery. In the lower portion of its course, it is accompanied by the anterior tibial nerve which lies to its outer side.

The **branches** of the anterior tibial artery are: (1) the *tibial recurrent*, (2) the *internal malleolar*, (3) the *external malleolar*, and (4) the *muscular*.

The **tibial recurrent artery** passes backward, between the popliteus muscle and the posterior ligament of the knee joint, to the anastomosis about the knee joint.

The **internal malleolar artery** assists in forming an anastomotic plexus in front of the internal malleolus.

The **external malleolar artery** goes to the external malleolus, over which it ramifies.

The **muscular branches** supply the muscles in the course of the anterior tibial artery.

The **dorsalis pedis artery** is the continuation of the anterior tibial artery, after that vessel reaches the inferior border of the anterior annular ligament. It extends to the first interosseous space where it bifurcates into its terminal branches.

The **branches** of the dorsalis pedis artery are: (1) the *tarsal*, (2) the *metatarsal*, (3) the *dorsalis hallucis*, and (4) the *communicating*.

The **tarsal branches** supply the tarsal joints, the tarsal bones, and the muscles with which they come in relation.

The **metatarsal artery** passes outward between the extensor brevis digitorum muscle and the metatarsal bones. The vessel describes a curve, the convexity of which is directed downward. From the convexity of this arch three **dorsal interosseous arteries** come off. These vessels pass downward in the interosseous spaces, and, at the web of the toes, divide into **dorsal digital** branches which supply the adjacent sides of the fifth and fourth, the fourth and third, and the third and second toes. The outer side of the fifth toe is supplied by a branch from the first dorsal interosseous artery. The inner side of the second toe and the dorsal aspect of the first toe are supplied by the dorsalis hallucis artery.

The **dorsalis hallucis artery** is a branch of the dorsalis pedis artery. It is given off in the first interosseous space just in front of the tarso-metatarsal articulation. It passes forward along the first interspace, and, at the web of the toes, divides into dorsal digital branches which supply the adjacent sides of the first and second toes. Before it divides, the dorsalis hallucis artery gives off a branch to the inner side of the first toe.

The **communicating artery** is a branch of the dorsalis pedis artery. It is given off in the first interosseous space, just in front of the tarso-metatarsal joint. It passes into the sole of the foot between the two heads of the first dorsal interosseous muscle, and joins with the external plantar artery to form the plantar arch. Just before the communicating artery joins with the external plantar artery, it gives off the **princeps hallucis artery** which is distributed to the inner side of the plantar aspect of the first toe and to the adjacent sides of the first and second toes. (Morris, p. 620; Gray, p. 641.)

The **posterior tibial artery** is a branch of the popliteal artery. It is given off at the lower border of the popliteus muscle and passes down the posterior aspect of the leg, resting on the tibialis posticus muscle, the flexor longus digitorum muscle and the posterior aspect of the tibia. Between the internal malleolus and the inner border of the os calcis it divides into its terminal branches. The posterior tibial nerve lies, first, to its inner side; but soon crosses above the vessel and lies to its outer side. These relations are given as the body is dissected. At the internal malleolus, the artery lies between the tendon of the flexor longus digitorum muscle and the posterior tibial nerve, having one of its venæ comites on either side.

The **branches** of the posterior tibial artery are: (1) the *muscular*, (2) the *internal calcanean*, (3) the *nutrient*, (4) the *communicating*, (5) the *peroneal*, (6) the *internal plantar*, and (7) the *external plantar*.

The **muscular branches** supply the muscles in the course of the artery.

The **internal calcanean artery** ramifies over the inner aspect of the os calcis.

The **nutrient artery** supplies the tibia.

The **communicating branches** join with the communicating branches of the peroneal artery to form a plexus around the tendo Achilles.

The **peroneal artery** passes down the outer border of the posterior aspect of the leg, resting on the fibula, and terminates by forming a dense network of fine branches over the external malleolus and the outer aspect of the os calcis. The **branches** of the peroneal artery are: (1) the *muscular*, (2) the *nutrient*, (3) the *communicating*, (4) the *anterior peroneal*, and (5) the *external calcanean*.

The **muscular branches** supply the muscles in the course of the vessel.

The **nutrient artery** supplies the fibula.

The **communicating branches** join with the communicating branches of the posterior tibial artery around the tendo Achilles.

The **anterior peroneal artery** is given off from the peroneal artery at the lower portion of the space between the tibia and the fibula. It pierces the interosseous membrane and passes over the inferior tibio-fibular articulation to terminate in a plexus covering the external malleolus.

The **external calcanean artery** is distributed to the outer aspect of the os calcis.

The **internal plantar artery** runs along the inner aspect of the foot, between the abductor hallucis and the flexor brevis digitorum muscles. It supplies the tissues of the sole of the foot and usually anastomoses with the princeps hallucis artery (see p. 168).

The **external plantar artery** passes beneath the abductor hallucis and the flexor brevis digitorum muscles, and then lies between the abductor minimi digiti and the flexor brevis digitorum muscles, until it reaches a point opposite the base of the fifth metatarsal bone. Here it sinks down deeply into the sole of the foot and passes across the metatarsal bones to join with the communicating branch of the dorsalis pedis artery to form the **plantar arch.** The external plantar artery gives cutaneous, muscular, and articular branches to the tissues with which it comes in relation. The **branches** of the plantar arch are: (1) the *perforating,* and (2) the *digital.* The **perforating arteries,** three in number, pass through the interosseous spaces and anastomose with the dorsal interosseous arteries, which come from the metatarsal artery. The **digital arteries,** four in number, supply the plantar surfaces of the toes. The first supplies the outer side of the little toe. The second, the third, and the fourth supply the adjacent sides of the fifth and fourth, the fourth and third, and the third and second toes, respectively. The adjacent sides of the second and first toes and the inner side of the first toe are supplied by the princeps hallucis artery, a branch of the communicating branch of the dorsalis pedis. (Morris, p. 614; Gray, p. 644.)

THE SACRAL PLEXUS.

The **sacral plexus** is formed by the lumbo-sacral cord and the anterior divisions of the first four sacral nerves. The **lumbo-sacral cord** is formed by the anterior division of the fifth lumbar nerve and a branch from the anterior division of the fourth lumbar nerve. The sacral plexus is found in the true pelvis, resting upon the pyriformis muscle.

The **branches** of the sacral plexus are: (1) the *superior gluteal,* (2) the *inferior gluteal,* (3) the *visceral,* (4) the *muscular,* (5) the *internal pudic,* (6) the *small sciatic,* and (7) the *great sciatic.*

The **superior gluteal nerve** leaves the pelvis by passing through the great sacro-sciatic foramen, above the tendon of the pyriformis muscle, in company with the gluteal artery. It supplies the gluteus medius, the gluteus minimus, and the tensor vaginæ femoris muscles.

The **inferior gluteal nerve** leaves the pelvis by passing through the great sacro-sciatic foramen below the tendon of the pyriformis muscle. It supplies the gluteus maximus muscle.

The **visceral nerves** pass to be distributed to the pelvic viscera.

The **muscular branches** supply the pyriformis, the obturator internus, the superior gemellus, the inferior gemellus, and the quadratus femoris muscles.

The **internal pudic nerve** leaves the pelvis by passing through the great sacro-sciatic foramen, below the tendon of the pyriformis muscle. It passes, in company with the internal pudic artery, across the spine of the ischium, through the lesser sacro-sciatic foramen, into the ischio-rectal fossa. In the ischio-rectal fossa it lies in Alcock's canal; it then pierces the superior layer of the triangular ligament, its terminal branch being found in the deep perineal interspace. The **branches** of the internal pudic nerve are: (1) the *inferior hemorrhoidal*, (2) the *perineal*, and (3) the *dorsal nerve of the penis.*

The **inferior hemorrhoidal nerve** is given off in Alcock's canal; it supplies the sphincter ani muscle and the skin around the anus.

The **perineal nerve** supplies the skin of the perineum and sends branches to the accelerator urinæ (sphincter vaginæ), the transversus perinei, the erector penis (erector clitoridis), and the compressor urethræ muscles. It is given off in Alcock's canal.

The **dorsal nerve of the penis** pierces the inferior layer of the triangular ligament. It sends branches to the corpus cavernosum and then passes down the dorsal aspect of the penis, sending branches to the prepuce and to the glans penis. In the female this nerve is much smaller than it is in the male. It is known as the **dorsal nerve of the clitoris.**

The **small sciatic nerve** leaves the pelvis by passing through the great sacro-sciatic foramen, below the tendon of the pyriformis muscle. It passes down the back of the thigh beneath the fascia lata. It pierces the deep fascia of the leg a short distance below the knee and ends at about the middle of the posterior aspect of the leg. In its course it gives a branch to the gluteus maximus muscle, branches to the skin of the thigh and leg, and the *inferior pudendal nerve.* The **inferior pudendal nerve** passes, in the gluteo-femoral fold, to the perineum. It sends branches to the scrotum, in the male, and to the labium major, in the female. (Morris, p. 834; Gray, p. 859.)

THE GREAT SCIATIC NERVE.

The **great sciatic nerve** is a branch of the sacral plexus. It leaves the pelvis by passing through the great sacro-sciatic foramen, below the tendon of the pyriformis muscle. It passes down the thigh resting on the gemellus superior, the obturator internus, the gemellus inferior, the quadratus femoris, and the adductor magnus muscles. In the upper part of the popliteal space it divides into the *external popliteal* and the *internal popliteal nerves.* In its course, it gives branches to the biceps, the semimembranosus, the semitendinosus, and the adductor magnus muscles.

The **external popliteal nerve** passes diagonally outward, lying internal to the tendon of the biceps muscle. It then passes between the peroneus longus muscle and the head of the fibula, where it divides into its terminal branches. The **branches** of the external popliteal nerve are: (1) the *articular*, (2) the *cutaneous*, (3) the *musculo-cutaneous*, and (4) the *anterior tibial.*

The **articular branches** supply the knee joint.

The **cutaneous branches** are distributed to the skin of the leg. One of these nerves joins with a branch from the internal popliteal nerve to form the external saphenous nerve. It is known as the **communicans peronei** or the **communicans fibularis.**

The **musculo-cutaneous nerve** is a branch of the external popliteal nerve. It passes through the substance of the peroneus longus muscle, sending branches to it and to the peroneus brevis muscle. It then pierces the deep fascia of the leg, at the junction of its middle and lower thirds, and becomes an occupant of the superficial fascia. It passes above the anterior annular ligament of the ankle and supplies the skin on the dorsum of the foot. It also supplies the dorsal aspects of the inner side of the first toe and the adjacent sides of the second and third, the third and fourth, and the fourth and fifth toes.

The **anterior tibial nerve** is a branch of the external

popliteal nerve. It pierces the extensor longus digitorum muscle and then passes down the leg, lying to the outer side of the anterior tibial artery, and between the tibialis anticus and the extensor longus digitorum muscles, the tibialis anticus and extensor proprius hallucis muscles, and the extensor proprius hallucis and the extensor longus digitorum muscles. It passes beneath the anterior annular ligament into the foot. It supplies the tibialis anticus, the extensor longus digitorum, the extensor proprius hallucis and the extensor brevis digitorum muscles; it sends branches to the ankle joint and to the tarsal joints; and it supplies the adjacent surfaces of the first and second toes.

The **internal popliteal nerve** passes through the middle of the popliteal space, lying superficial to and somewhat more external than the popliteal vessels. As the nerve passes through the popliteal space it crosses above the popliteal vessels and then lies internal to them. At the lower border of the popliteus muscle it becomes the posterior tibial nerve. It gives off a branch, the **communicans popletei,** which joins with the communicans peronei branch of the external popliteal nerve to form the *short saphenous nerve.* The **short saphenous nerve** passes down the posterior aspect of the leg, supplying the skin in that region, and along the outer border of the foot. It ends by supplying the outer side of the fifth toe. In its course it is accompanied by the short saphenous vein. The internal popliteal nerve sends branches to the gastrocnemius, the soleus, the plantaris, and the popliteus muscles. It also gives filaments to the knee joint.

The **posterior tibial nerve** is the continuation of the internal popliteal nerve, from the lower border of the popliteus muscle. It passes down the leg resting on the tibialis posticus, and the flexor longus digitorum muscles. In the upper part of the leg the posterior tibial nerve lies to the inner side of the posterior tibial artery; but it soon crosses above the artery and then is found to its outer side. It passes behind the internal malleolus, lying between the tendon of the flexor longus hallucis muscle and the posterior tibial vessels. Just below the internal malleolus it divides into its terminal branches. The posterior tibial nerve supplies the tibialis posticus, the flexor longus digitorum, and the flexor longus hallucis muscles. It gives a branch to the skin of the sole of the foot and divides into the *external* and the *internal plantar nerves.*

The **external plantar nerve** passes between the flexor brevis digitorum and the flexor accessorius muscles and supplies the muscles on the outer side of the sole of the foot and the plantar surfaces of the fifth and the outer half of the fourth toes.

The **internal plantar nerve** passes between the abductor hallucis and the flexor brevis digitorum muscles. It supplies the muscles on the inner side of the sole of the foot and the plantar surfaces of the first, second, and third toes and the inner half of the fourth toe. (Morris, p. 838; Gray, p. 862.)

The **cutaneous nerve supply of the buttock** is derived from the iliac branch of the ilio-hypogastric nerve, the terminal branch of the last dorsal nerve, branches of the posterior divisions of the sacral and the lumbar nerves, and a branch from the inferior gluteal nerve.

The **cutaneous nerve supply of the thigh** is as follows: on the anterior surface, the crural branch of the genito-crural, the inguinal branch of the ilio-inguinal, the external cutaneous, the middle cutaneous, and the internal cutaneous nerves. On the dorsal surface, the posterior branch of the external cutaneous nerve, the posterior branch of the internal cutaneous nerve, and branches of the small sciatic nerve.

The **cutaneous nerve supply of the leg** is as follows: on the anterior surface, the long saphenous nerve, a branch of the external popliteal nerve, and branches of the musculo-cutaneous nerve. On the posterior surface, the small sciatic nerve, the short saphenous nerve, and branches of the external popliteal nerve. Twigs from the long saphenous nerve also pass over to the posterior aspect of the leg.

The **cutaneous nerve supply of the foot** is as follows: on the dorsal surface, the long saphenous nerve to the ball of the great toe, the anterior tibial nerve to the adjacent sides of the first and second toes, the short saphenous nerve to the outer side of the fifth toe, and the musculo-cutaneous nerve to the inner side of the first toe, and to the adjacent sides of the second and third, the third and fourth, and the fourth and fifth toes. On the plantar surface, the internal plantar nerve to the first, second, and third toes, and to the inner side of the fourth toe, and the external plantar nerve to the fifth toe and to the outer side of the fourth toe. The plantar cutaneous branch of the posterior tibial nerve is distributed to the posterior part of the plantar surface of the foot.

CHAPTER XI.

THE PELVIC OUTLET.

The **outlet of the pelvis** is a quadrilateral figure which is bounded, *anteriorly*, by the rami of the pubes and ischium, on either side; and *posteriorly*, by the sacro-sciatic ligaments, on either side. The anterior angle of this figure is formed by the subpubic arch; the posterior angle is occupied by the tip of the coccyx; and the lateral angles correspond to the tuberosities of the ischia. This opening is closed in by muscles and fascias through which the rectum and the urethra pass. In the female, the vagina, in addition to the canals just mentioned, perforates the floor of the pelvis.

The fascial tissues which close in the pelvic outlet are continuous with the **iliac fascia,** which covers the iliacus muscle. At the ilio-pectineal line, the iliac fascia passes downward into the true pelvis and is then known as the **pelvic fascia.** At the **white line,** which extends from the symphisis pubis to the spine of the ischium, the pelvic fascia divides into the *recto-vesical fascia* and the *obturator fascia.* The **recto-vesical fascia** passes inward toward the midline and gives off the **rectal fascia,** which surrounds the rectum, and the **vesical fascia,** which envelops the bladder and sends a process to surround the prostate gland. In the female the recto-vesical fascia ensheaths the uterus and the vagina. The **obturator fascia** passes downward along the bony wall of the pelvis, covering in the obturator internus muscle. At the margin of the pelvic outlet this fascia sends off a triangular layer of tissue which extends between the rami of the pubes and ischium, from the sub-pubic arch to the tuberosities of the ischia; this is known as the **superior layer of the triangular ligament.** The recto-vesical fascia, after enveloping the prostate gland, contributes to the formation of the superior layer of the triangular ligament. A second process, the **ischio-rectal fascia** or

176

the **anal fascia,** passes backward, lining the ischio-rectal fossa. The levator ani muscle is covered on the superior and internal surface by the recto-vesical fascia, on its outer aspect by the obturator fascia, and on its inferior surface by the anal fascia. The levator ani muscle, the recto-vesical fascia, and the coccygeus muscle form the true floor of the pelvis.

When studied from the cutaneous aspect, the region of the pelvic outlet may be divided by a line drawn between the tuberosities of the ischia, into an anterior, **urethral triangle** and a posterior, **anal triangle.** (Morris, p. 1072; Gray, p. 1201.)

THE URETHRAL TRIANGLE.

The urethral triangle is bounded, *behind,* by a line drawn between the tuberosities of the ischia, and *on either side,* by the rami of the pubes and ischium. This triangle is the **perineum.** It is pierced by the urethra, and, in the female, by the vagina.

The **superficial fascia** of the perineum is divisible into two layers. The superficial layer of the superficial fascia of the perineum is continuous with the superficial fascia of the thighs and with the dartos of the scrotum. The deep layer of the superficial fascia of the perineum is attached on either side to the rami of the pubes and of the ischium, and dips posteriorly to be attached to the deep perineal fascia. This tissue is called **Colles' fascia.**

The **deep fascia** of the perineum is spoken of as the **inferior layer of the triangular ligament.** It is attached to the rami of the pubes and ischium and passes upward to be attached to the superior layer of the triangular ligament, which is formed from the obturator and recto-vesical fascias (see p. 176). It also receives the posterior end of Colles' fascia.

Between Colles' fascia and the inferior layer of the triangular ligament there is a space which is known as the **superficial perineal interspace.** This space contains the bulbous urethra, the accelerator urinæ muscle (sphincter vaginæ), the erector penis muscle (erector clitordis), the transversus perinei muscle, the superficial perineal vessels and nerves, the arteries of the corpora cavernosa, the dorsal arteries of the penis, and the dorsal nerves of the penis.

Between the inferior layer of the triangular ligament and the superior layer of the triangular ligament we find the **deep perineal interspace.** It contains the membranous urethra, the compressor urethræ muscle, the internal pudic arteries, the dorsal arteries of the penis, the arteries to the bulb, the arteries to the corpora cavernosa, the internal pudic nerves, the dorsal nerves of the penis, and Cowper's glands.

In the female, the tissue situated between the anus and the posterior boundary of the vulva is spoken of as the **perineum** or the **perineal body.** (Morris, p. 1080; Gray, p. 1202.)

THE ANAL TRIANGLE.

The **anal triangle** is bounded, *in front,* by a line drawn between the tuberosities of the ischia; and on *either side posteriorly,* by the sacro-sciatic ligaments or by the margins of the two glutei maximi muscles. It contains the anus, surrounded by the *sphincter ani muscle,* the *inferior hemorrhoidal vessels* and *nerves,* and the *ischio-rectal fossæ.*

The **ischio-rectal fossa** is the space between the wall of the rectum and the ischium. It is bounded, *externally,* by the obturator fascia, covering the obturator internus muscle; *internally,* by the anal fascia, covering the levator ani muscle; and *below,* by the fascias and skin of the anal triangle. It contains much *fat,* the *inferior hemorrhoidal arteries* and *nerves,* and the *internal pudic arteries* and *nerves.* The internal pudic artery and the internal pudic nerve lie in Alcock's canal, which is a splitting of the obturator fascia on the outer wall of the space. (Morris, p. 1079; Gray, p. 1201.)

CHAPTER XII.

THE LYMPHATIC SYSTEM.

The **lymphatic system** is composed of the lymphatic vessels and the lymphatic tissues.

The **lymphatic vessels** may be classified as; first, undefined spaces, seen in the connective tissues, as the juice canals; second, anastomosing clefts; and third, the distinct vessels. The **lymphatic tissue** is of three varieties; first, the diffuse adenoid tissue; second, simple follicles, such as the solitary follicles in the wall of the intestine; and third, compound lymph nodes, such as the inguinal and the axillary lymphatics.

The lymphatic vessels are, for the most part, composed of a single layer of endothelial cells. The larger vessels, such as the thoracic duct, present an intima, a media, and an adventitia. The lymphatics are furnished with valves. Throughout the body the lymphatic vessels may be divided into a superficial group and a deep group.

The **superficial lymphatic vessels of the head and neck** empty into lymphatic glands which are situated behind the ear, **posterior auricular glands;** beneath the lower jaw, **submaxillary glands;** in front of the parotid gland, **parotid lymphatics;** and along the course of the external jugular vein. The **deep lymphatic glands** are situated, principally, along the course of the internal jugular vein. The lymphatic vessels from the mouth and tongue empty partly into the submaxillary glands and partly into the deep cervical glands.

The **lymphatic vessels of the upper extremity** begin at the finger tips. They pass along the dorsal and palmar surfaces of the hand, forearm, and arm to empty into the axillary lymph glands. The **axillary glands** are arranged in four groups. The first group, the **axillary glands,** follows the course of the axillary vein. The second group, the **pectoral glands,** is found along the long thoracic artery. The third group, the **subscapular glands**, lies parallel with the subscapular artery. The fourth group, the **infraclavicular glands,** lies below the clavicle and on the costo-coracoid membrane. The lymphatic vessels from the mammary gland empty, principally, into the axillary glands and partly into the mediastinal glands.

The **lymphatic vessels of the lower extremity** begin in a close network at the ends of the toes. These vessels pass upward, following the courses of the saphenous veins, to empty into the inguinal glands. The **inguinal** glands are arranged in two groups. The **inguinal glands** lie parallel to Poupart's ligament; they receive the lymphatic vessels from the penis and from the abdominal walls. The **saphenous glands** are arranged about the saphenous opening; they receive the lymphatic vessels from the foot and leg. The **popliteal glands** follow the course of the popliteal vessels. The **deep femoral glands** are in relation with the femoral vessels in the upper part of Scarpa's triangle. The femoral canal usually holds one of these glands.

The **lymphatic vessels of the intestine** begin in the villi, as the lacteals. They pass through the lymph nodes in the intestinal wall and through the **mesenteric glands** to empty, finally, into the receptaculum chyli.

The **lymphatic vessels of the liver** empty partly into the mediastinal glands, partly into the glands in the lesser omentum, and partly into the receptaculum chyli, or into the lumbar glands.

The abdomen and pelvis contain groups of lymphatic glands, the situation of which is indicated by their names. They are the **external iliac glands,** the **internal iliac glands,** the **sacral glands,** the **lumbar glands,** the **celiac glands,** the **gastric glands,** the **mesenteric glands,** the **mesocolic glands,** the **hepatic glands** and the **splenic glands.**

The lymphatics of both lower extremities, of the abdomen, the pelvis, the thorax, the left upper extremity, and the left side of the head empty into the thoracic duct. The **thoracic duct** begins in a dilated extremity or pouch which is known as the receptaculum chyli. The **receptaculum chyli** is situated on the body of the second lumbar vertebra, between and behind the abdominal aorta and the inferior vena cava. The thoracic duct begins at the upper margin of the receptaculum chyli. It passes through the aortic opening in the diaphragm, into the thorax, lying between and behind the aorta and the vena azygos major. It passes upward through the posterior mediastinum, having the same relation with the aorta and the vena azygos major. At the lower border of the fourth thoracic vertebra it enters the superior mediastinum and, passing through this space, lying behind the arch of the aorta and the left subclavian artery, it enters the root of the neck, and empties into the junction of the left subclavian and the left internal jugular veins. The thoracic duct is about eighteen inches long.

The **right lymphatic duct** receives the lymphatics from the right arm and from the right side of the head. It empties into the junction of the right subclavian and the right internal jugular veins. It is about one-half inch in length. (Morris, p. 673; Gray, p. 679.)

CHAPTER XIII.

THE HEART AND THE GREAT BLOOD VESSELS.

The heart is a hollow muscular organ, which is contained in the middle mediastinum. It is surrounded by the pericardium.

The **pericardium** is the serous membrane which surrounds the heart. It is pyramidal in shape, being attached by its base to the central tendon of the diaphragm, and by its apex to the ascending portion of the arch of the aorta. It is composed of two layers; a **parietal layer,** which forms the walls of the closed sac which contains the heart; and a **visceral layer,** which is reflected along the blood vessels and which closely invests the heart itself. Histologically, the pericardium is composed of a **serous layer,** which lines the cavity, and of a **fibrous layer,** which is continuous with the pretracheal fascia in the neck. The cavity of the pericardium contains a straw colored serum which is normal in amount up to 100 cubic centimetres. Traversing the cavity of the pericardium, on their way to and from the heart, we find the aorta, the pulmonary artery, the superior vena cava, and the four pulmonary veins. The **oblique sinus of the pericardium** is the space between the inferior vena cava and the left inferior pulmonary vein. The **vestigial fold of the pericardium** is a triangular fold of the serous layer, which passes from the pulmonary artery to the left superior pulmonary vein. It contains the obliterated remains of the left superior vena cava. The pericardium is attached to the sternum by the **sterno-pericardiac ligaments.** (Morris, p. 941; Gray, p. 1083.)

THE HEART.

The **heart** is conical in shape and is placed obliquely in the thorax with its base upward and its apex downward and to the left. A line drawn from the upper border of the third costal cartilage, one-half inch to the right of the sternum, obliquely across the second interspace, to the lower border of the second costal cartilage, one inch to the left of the sternum, would represent the position of the base of the organ. The apex is situated in the fifth left intercostal space, about three and one-half inches to the left of the midsternal line. A line connecting the left end of the line of the base of the heart with the position of the apex would represent the left border of the organ. The right border of the heart would be represented by a line drawn from the right extremity of the base line to the position of the apex.

On examining the external wall of the heart it will be seen to be divided into a superior, auricular portion and an inferior ventricular portion by a transverse groove, the **auriculo-ventricular groove.** The **interauricular groove** is seen on the auricular portion dividing it into a left segment and a right segment. The **interventricular groove,** similarly, divides the ventricular portion into a right and a left segment. These grooves are filled by a varying amount of fat.

The heart weighs about nine ounces (270 grams) in the the female, and about eleven ounces (330 grams) in the male.

When examined from its inner aspect the heart is found to present four cavities for study; a right and a left auricle above; and a right and a left ventricle below. The auricles are separated from the ventricles by the **auriculo-ventricular septum.** The auricles are separated from each other by the **interauricular septum.** The ventricles are separated from each other by the **interventricular septum.** These septa correspond in position to the grooves seen on the external surface of the organ. The right auricle opens into the right ventricle through the **right auriculo-ventricular opening.** The left auricle communicates with the left ventricle by means of the **left auriculo-ventricular** opening. In the adult, there is, normally, no communication between the two auricles and none between the two ventricles. In the fetus, the right auricle communicates with the left auricle through the **foramen ovale.** Each auricle presents a small dilatation of its cavity known as the **right** and the **left auricular appendix,** respectively.   The cavity of the right auricle is often called the **sinus venosus.**

The heart is lined by a delicate, serous membrane, which is known as the **endocardium.**

The **right auricle** presents the following structures for examination: (1) the *opening of the superior vena cava,* (2) the *opening of the inferior vena cava,* (3) the *opening of the coronary sinus,* (4) the *foramina Thebesii,* (5) the *auriclo-ventricular opening,* (6) the *fossa ovalis,* (7) the *annulus ovalis,* (8) the *Eustachian valve,* (9) the *tubercle of Lower,* and (10) the *musculi pectinati.*

The **foramina Thebesii** are the openings of small veins which return blood from the wall of the auricle to the cavity of the right auricle.   Some of these openings are blind.

The **fossa ovalis** is the remains of the fetal foramen ovale.

The **annulus ovalis** is the prominent margin of the fossa ovalis.

The **Eustachian valve** is a fold of endocardium which passes from the opening of the inferior vena cava to the fossa ovalis.   In the fetus, it serves to direct the blood through the foramen ovale.

The **tubercle of Lower** is a rounded prominence between the orifices of the two venæ cavæ.   In the fetus, it directs the blood from the superior vena cava to the auriculo-ventricular opening.

The **musculi pectinati** are raised bands of muscular tissue which are found on the anterior wall of the auricle and in the auricular appendix.

The **right ventricle** presents for examination: (1) the *tricuspid valve,* (2) the *orifice of the pulmonary artery,* (3) the *pulmonary semilunar valves,* (4) the *corpora Arantii,* (5) the *sinuses of Valsalva,* (6) the *columnæ carneæ,* (7) the *musculi papillares,* and (8) the *chordæ tendineæ.*

The **tricuspid valve** guards the right auriculo-ventricular opening.   It is composed of three leaflets, which are formed by folds of endocardium.   These leaflets are attached by their bases to the auriculo-ventricular septum. The free margins of the valve project into the ventricular cavity, and are connected to the apices of the musculi papillares by numerous fibrous bands, the **chordæ tendineæ.**

The **pulmonary artery** arises from a portion of the cavity of the right ventricle which is known as the **conus arteriosus**.

The **pulmonary semilunar valves** are three, semilunar folds of endocardium which are attached by their bases to the fibrous ring from which the artery takes origin. The free edges of these valves are known as the **lunulæ**. Two of these leaflets are placed anteriorly and one is placed posteriorly.

The **corpora Arantii** are small fibrous nodules in the centre of the lunulæ.

The **sinuses of Valsalva** are small pouches between the semilunar valve leaflets and the wall of the artery.

The **columnæ carneæ** are muscular ridges which are seen on the wall of the ventricle, producing an extremely rough appearance. These muscular ridges are of three kinds; first, a group, the members of which are attached to the wall of the ventricle for their entire extent, merely producing a ridge on the wall of the ventricle; second, a group, the members of which are attached to the wall of the ventricle by either end, leaving a free space under the middle of the band, beneath which a probe may be passed; and third, a group, the members of which are attached to the wall of the ventricle by one end only, the other end lying free in the cavity of the ventricle. The members of the third group are known as the **musculi papillares.** There are three musculi papillares in the right ventricle; one on the anterior wall; one on the interventricular septum, and one on the anterior wall near the right margin.

The **moderator band** is a band of muscular tissue which runs from the anterior wall of the right ventricle to the interventricular septum. It is not constant.

The **left auricle** presents for examination: (1) the *openings of the four pulmonary veins*, (2) the *left auriculo-ventricular opening*, and (3) the *musculi pectinati*.

The **musculi pectinati** are muscular ridges which are best

seen in the auricular appendix. They are not so well marked as
they are in the right auricle.

The **left ventricle presents** for examination : (1) the
mitral valve, (2) the *opening of the aorta,* (3) the *aortic semilunar valves,* (4) the *corpora Arantii,* (5) the *sinuses of Valsalva,*
(6) the *columnæ carneæ,* (7) the *musculi papillares,* and (8) the
chordæ tendineæ.

The **mitral valve** guards the left auriculo-ventricular opening, projecting into the left ventricle. It has two leaflets, which
are composed of a folding of the endocardium. The valve is
attached by its base to the auriculo-ventricular septum and
its free margin is connected with the musculi papillares by the
chordæ tendineæ. That leaflet of the mitral valve which
looks toward the beginning of the aorta is known as the **aortic
leaflet** of the mitral valve.

The **aortic semilunar valves** are three semilunar folds
of endocardium which are attached by their bases to the fibrous
ring from which the aorta springs. The free margins of these
folds are known as the **lunulæ.** Two of these leaflets are
placed posteriorly and one is placed anteriorly.

The **corpora Arantii** are small fibrous nodules in the
center of the lunulæ.

The **sinuses of Valsalva** are the pockets between the
leaflets of the semilunar valves and the wall of · the aorta.
The right coronary artery arises form the anterior sinus of
Valsalva ; the left coronary artery arises from the left posterior
sinus of Valsalva.

The **columnæ carneæ** and the **musculi papillares** are similar to those structures, bearing like names, described in the
right ventricle. There are usually but two musculi papillares
in the left ventricle, one on the anterior wall, and one on the
posterior wall of the ventricle. (Morris, p. 942; Gray, p. 1086.)

THE RELATION OF THE VALVES TO THE CHEST WALL.—
The **tricuspid valve** is situated in the midsternal line, opposite
the fourth costal cartilage. The **mitral valve** is situated in
the third intercostal space, about one inch to the left of the
sternum. The **aortic semilunar valves** are situated in the

third intercostal space at the left margin of the sternum. The
pulmonary semilunar valves are found behind the articulation of the third costal cartilage and the sternum.

The heart is supplied with blood by the right and the left
coronary arteries. The **right coronary artery** arises from the
right anterior sinus of Valsalva. It passes along the auriculo-
ventricular groove to the right and, on the posterior wall of the
organ, at the interventricular groove, divides into a branch which
passes down the interventricular groove, and a branch which
passes along the auriculo-ventricular groove to anastomose with the
left coronary artery. The **branches** of the right coronary artery
are: (1) the *auricular*, to the right auricle, (2) the *preventricular*,
to the anterior wall of the right ventricle, (3) the *right marginal*,
to the right margin of the heart, and (4) the *interventricular*,
in the posterior interventricular groove.

The **left coronary artery** arises from the posterior sinus
of Valsalva and passes forward in the auriculo-ventricular
groove. It sends one branch backward to anastomose with
the termination of the right coronary artery and another to the
apex of the heart in the anterior interventricular groove. The
branches of the left coronary artery are: (1) the *auricular*, to
the left auricle, (2) the *left marginal*, along the left margin of
the heart, and (3) the *interventricular* in the anterior interventricular groove.

The **coronary veins** pass with the coronary arteries. The
great cardiac vein passes upward in the anterior interventricular groove and then, running to the left in the auriculo-
ventricular groove, joins with the **posterior cardiac vein,**
which passes upward in the posterior interventricular groove, to
form the coronary sinus. The **coronary sinus** empties into
the right auricle.

The **cardiac nerves** are derived from the deep and the
superficial cardiac plexuses (see pp. 82 and 83). (Morris, p.
951; Gray, pp. 542 and 677.)

THE GREAT BLOOD VESSELS.—THE PULMONARY ARTERY.

The **pulmonary artery** arises from the conus arteriosus

of the right ventricle. It lies in front of the origin of the aorta and is overlapped, somewhat, by the left auricular appendix. The vessel passes to the left, across the cavity of the pericardium and beneath the transverse portion of the arch of the aorta. About two inches after the vessel is given off from the right ventricle it divides into the *right pulmonary artery* and the *left pulmonary artery.*

The **right pulmonary artery** passes to the right, beneath the arch of the aorta, to the root of the right lung. It gives branches to the three lobes of the right lung.

RELATIONS.—In front, the ascending portion of the arch of the aorta, the superior vena cava, and the phrenic nerve. Behind, the right bronchus. Above, the transverse portion of the arch of the aorta. Below, the right auricle.

The **left pulmonary artery** passes to the left and enters the root of the left lung, sending branches to the two lobes of that organ.

RELATIONS.—In front, the phrenic nerve. Behind, the descending portion of the arch of the aorta, the left pneumogastric nerve, and the left bronchus. Below, the pulmonary veins.

The obliterated remains of the ductus arteriosus pass from the left pulmonary artery to the aorta.

THE ARCH OF THE AORTA.

The **aorta** is divisible into the *arch of the aorta*, the *thoracic aorta*, and the *abdominal aorta.*

The **arch of the aorta** is situated in the superior mediastinum; it may be subdivided into, the **ascending portion,** which extends from the point of origin of the vessel from the heart to the second right costal cartilage; the **transverse portion,** which passes transversely across the middle of the first piece of the sternum to the left side of the body of the fourth thoracic vertebra; and the **descending portion,** which extends from the upper border of the fourth to the lower border of the fifth thoracic vertebra. In its course it describes a curve from before backward, as well as from right to left. The arch of the aorta does not present the same calibre throughout its entire extent; but, on the contrary, shows certain dilatations and constrictions of its lumen. The **great sinus** is a dilation of the right wall of the aorta, due to the force of the blood which strikes it at each pulsation of the heart. The **aortic isthmus** is the most contracted portion of the vessel; it is seen just below the point of origin of the left subclavian artery. The **aortic spindle** is the name given to that portion of the arch of the aorta beyond the aortic isthmus.

RELATIONS.—The ascending portion of the arch of the aorta is in relation with the following structures: in front, the right auricular appendix, the pulmonary artery, the pericardium, the remains of the thymus gland, and the right pleura; behind, the left auricle, the right pulmonary artery, and the right bronchus; to the right, the right auricle and the superior vena cava; to the left, the pulmonary artery. The transverse portion of the arch of the aorta is in relation with the following structures: in front, the right and the left pleuræ, the left phrenic nerve, the left pneumogastric nerve, the left superior intercostal vein, and the cardiac nerves; behind, the trachea, the esophagus, the thoracic duct, the left recurrent laryngeal nerve, and the deep cardiac plexus; above, the left innominate vein, and the origins of the innominate, the left common carotid, and the left subclavian arteries; below, the pulmonary artery, the superficial cardiac plexus, the left bronchus, and the left recurrent laryngeal nerve. The descending portion of the arch of the aorta is in relation with the following structures: in front, the left pleura and the root of the left lung; behind, the bodies of the fourth and fifth thoracic vertebræ; to the right, the esophagus, and the thoracic duct; to the left, the left pleura and the lung.

The **branches** of the arch of the aorta are: (1) the *innominate*, (2) the *left common carotid*, and (3) the *left subclavian*.

The **innominate** artery passes from the transverse portion of the arch of the aorta, obliquely across the superior mediastinum, to the right sterno-clavicular articulation, where it divides into the right common carotid artery and the right subclavian artery. (See pp. 110 and 116.)

RELATIONS.—In front, the thymus gland, the left innominate vein, and the inferior thyroid veins. Behind, the trachea and the right pleura. To the right, the right innominate vein and the right pneumogastric nerve. To the left, the left common carotid artery, the inferior thyroid veins, and the trachea. (Morris, p. 487; Gray, p. 543.)

For the description of the **left common carotid artery** see page 116.

For the description of the **left subclavian artery** see page 111.

THE THORACIC AORTA.

The **thoracic aorta** begins at the lower border of the fifth thoracic vertebra and passes downward, through the posterior mediastinum, to the aortic opening in the diaphragm, through which it passes, to become the abdominal aorta. As it passes through the posterior mediastinum it lies a little to the left of the median line.

RELATIONS.—In front, the root of the left lung, the esophagus, and the pericardium. Behind, the anterior surfaces of the bodies of the thoracic vertebræ, from the sixth to the twelfth, inclusive, the vena azygos minor, and the left upper azygos vein. To the right, the thoracic duct, the vena azygos major, and the right pleura. In the upper portion of its course the esophagus lies to its right side. To the left, the left pleura, the esophagus, the vena azygos minor and the left upper azygos vein.

The **branches** of the thoracic aorta are: (1) the *pericardiac*, (2) the *esophageal*, (3) the *mediastinal*, (4) the *bronchial*, and (5) the *intercostal*.

The **pericardiac branches** supply the pericardium.

The **esophageal branches** supply the esophagus.

The **mediastinal branches** supply the tissue in the posterior mediastinum.

The **bronchial arteries** pass along the posterior aspects of the bronchi and supply the lungs with nutrient blood. There are two left bronchial arteries and one right bronchial artery.

The **intercostal arteries,** ten pairs, supply the inter-costal spaces below the second. The vessels come off from the lateral aspects of the thoracic aorta and pass outward, across the bodies of the thoracic vertebræ, to the intercostal spaces. The right vessels are longer than the left, on account of the position of the thoracic aorta, to the left of the median line. As the arteries pass outward in the intercostal spaces they lie midway between the two ribs, as far as their angles; they then pass in a groove on the lower border of the upper rib, the subcostal groove. They end by anastomosing with the anterior intercostal branches of the internal mammary artery and the anterior intercostal branches of the musculo-phrenic artery. The intercostal arteries give branches to the muscles of the back, to the vertebræ, to the spinal cord, to the pleura, to the intercostal muscles, and to the skin of the thorax. The last intercostal artery passes forward along the lower border of the twelfth rib, and is sometimes called the **subcostal artery.** The lower intercostal arteries help to supply the anterior abdominal walls. (Morris, p. 568; Gray, p. 605.)

THE ABDOMINAL AORTA.

The **abdominal aorta** begins at the lower margin of the aortic opening in the diaphragm; it passes along the posterior abdominal wall, to the lower border of the fourth lumbar verte-bra, where it divides into its terminal branches. As it passes through the abdomen it lies to the left of the median line.

RELATIONS.—In front, the right lobe of the liver, the solar plexus, the lesser omentum, the esophagus, the stomach, the superior layer of the transverse mesocolon, the splenic vein, the pancreas, the left renal vein, the transverse portion of the duodenum, the mesentery, and the small intestine. Behind, the bodies the first four lumbar vertebræ, the left crus of the diaphragm, and the left lumbar veins. To the right, the inferior vena cava, the right crus of the diaphragm the great splanchnic nerve, the receptaculum chyli and the thoracic duct. To the left, the left crus of the diaphragm and the left splanchnic nerve.

The **branches** of the abdominal aorta are: (1) the *phrenics*, (2) the *celiac axis*, (3) the *suprarenals*, (4) the *renals*, (5) the *superior mesenteric*, (6) the *spermatics*, (7) the *inferior mesenteric*, (8) the *lumbar*, (9) the *middle sacral*, and (10) the *common iliacs*.

The **phrenic arteries,** two in number, pass upward and backward to supply the diaphragm.

The **celiac axis** comes off from the anterior aspect of the abdominal aorta. It is about one-half inch long.

RELATIONS.—In front, the lesser omentum. Behind, the aorta. To the right, the right semilunar ganglion and the lobus Spigelii of the liver. To the left, the left semilunar ganglion and the cardiac end of the stomach. Above, the right lobe of the liver. Below, the pancreas.

The **branches** of the celiac axis are: (1) the *gastric*, (2) the *hepatic*, and (3) the *splenic*.

The **gastric artery** passes to the left end of the lesser curvature of the stomach and then passes along the lesser curvature, from left to right, supplying that region of the stomach. It anastomoses with the pyloric branch of the hepatic artery.

The **hepatic artery** is a branch of the celiac axis. It passes directly forward, until it reaches the lesser omentum, and then runs upward in that membrane to enter the transverse fissure of the liver. In its course through the lesser omentum, the common bile duct lies to the right of it and the portal vein lies between and behind it and the bile duct.

The **branches** of the hepatic artery are: (1) the *pyloric*, (2) the *cystic*, (3) the *gastro-duodenal*, (4) the *right terminal*, and (5) the *left terminal*.

The **pyloric artery** passes to supply the pyloric end of the stomach.

The **cystic artery** supplies the gall bladder.

The **gastro-duodenal** artery passes downward behind the pyloric end of the stomach, and divides into the **gastro-epiploica dextra artery,** which supplies the right side of the greater curvature of the stomach, and the **superior pan-creatico-duodenal artery**, which sends branches to the pancreas and to the duodenum. The gastro-epiploica dextra artery is found between the layers of the great omentum.

The **right** and the **left terminal branches** supply the right and the left lobes of the liver, respectively. They enter the organ by passing through the transverse fissure.

The **splenic artery** passes along the upper border of the pancreas and through the phreno-splenic ligament to supply the spleen. In its course it crosses over the right kidney.

The **branches** of the splenic artery are: (1) the *small pancreatic*, (2) the *large pancreatic*, (3) the *gastro-epiploica sinistra*, (4) the *vasa brevia,* and (5) the *terminal.*

The **small pancreatic arteries** supply the pancreas.

The **large pancreatic artery,** larger than the other pancreatic vessels, is distributed to the pancreas.

The **gastro-epiploica sinistra artery** passes between the layers of the great omentum and supplies the left side of the greater curvature of the stomach.

The **vasa brevia** pass through the gastro-splenic omentum to the fundus of the stomach.

The **terminal arteries** enter the spleen by passing through the hilum of the organ.

The **suprarenal arteries,** two in number, supply the right and the left suprarenal bodies. The suprarenal bodies also receive branches from the phrenic arteries and from the renal arteries.

The **renal arteries** are given off from the abdominal aorta a short distance below the superior mesenteric artery. The right renal artery passes beneath the inferior vena cava, in order to reach the right kidney. Each vessel enters the organ which it is to supply by passing through the hilum into the sinus.

The **superior mesenteric artery** comes off from the anterior aspect of the abdominal aorta. It passes between the lower border of the pancreas and the transverse portion of the duodenum and enters the mesentery. It supplies the small intestine and the ascending and transverse portions of the large intestine.

The **branches** of the superior mesenteric artery are: (1) the *inferior pancreatico-duodenal,* (2) the *ileo-colic,* (3) the *right colic,* (4) the *middle colic,* and (5) the *vasa intestini tenuis.*

The **inferior pancreatico-duodenal artery** passes backward to supply the pancreas and the duodenum.

The **ileo-colic artery** supplies the ileo-cecal region.

The **right colic artery** supplies the ascending colon. It anastomoses with the middle colic and with the ileo-colic arteries. It is found in the ascending mesocolon.

The **middle colic artery** is found in the transverse mesocolon. It anastomoses with the right colic and the left colic arteries and supplies the transverse colon.

The **vasa intestini tenuis,** ten or twelve in number, run in the mesentery. Each artery divides into two branches. These two branches are connected by an anastomotic arch, from the convex side of which other branches are given off. The secondary branches are connected by anastomotic arches and frequently, three or four of these arches, will be found between the main trunk of the superior mesenteric artery and the mesenteric attachment of the bowel. The final branches from the vascular arches, when they reach the intestine, divide into two branches, of which one passes behind the gut and the other passes in front of the gut, the two vessels anastomosing over the free margin of the bowel.

The **spermatic arteries,** two in number, arise from the anterior aspect of the abdominal aorta. They pass diagonally outward, across the psoas magnus muscle of either side, to the corresponding internal abdominal ring. The vessel then passes through the inguinal canal, as one of the constituents of the spermatic cord, to supply the epididymis and the testicle. In its course, the spermatic artery passes above the ureter. The **ovarian artery** in the female is the homologue of the spermatic artery. It passes over the psoas magnus muscle until it reaches the common iliac artery. It then passes through the infundibulopelvic ligament to the broad ligament. It passes between the layers of the broad ligament to supply the ovary and the Fallopian tube.

The **inferior mesenteric artery** is given off from the anterior aspect of the abdominal aorta. It passes across the left psoas magnus muscle and across the left common iliac artery, to break up into its terminal branches in the descending meso-colon.

The **branches** of the inferior mesenteric artery are: (1) the *left colic*, (2) the *sigmoid*, and (3) the *superior hemorrhoidal.*

The **left colic artery** runs in the descending mesocolon and supplies the descending colon. It anastomoses with the middle colic and the sigmoid arteries.

The **sigmoid artery** supplies the sigmoid flexure of the colon. It is found in the mesosigmoid.

The **superior hemorrhoidal artery** passes between the layers of the mesorectum, to the upper portion of the rectum.

The **lumbar arteries,** four pairs, are given off from the lateral aspects of the aorta. They pass outward, between the transverse processes of the lumbar vertebræ, to supply the tissues forming the abdominal walls.

The **middle sacral artery** is given off from the abdominal aorta at its point of bifurcation. It passes beneath the left common iliac vein, over the promontory of the sacrum, along the middle of the curve of the sacrum, to end in front of the coccyx by anastomosing with the lateral sacral arteries.

THE COMMON ILIAC ARTERIES.

The **common iliac arteries,** two in number, are the terminal branches of the abdominal aorta. They are given off opposite the lower border of the fourth lumbar vertebra and pass outward to the right and left sacro-iliac articulations, where each divides into the external iliac artery and the internal iliac artery.

RELATIONS.—The right common iliac artery is in relation with the following structures: in front, the peritoneum, the right ureter, the ileum, and the ovarian artery in the female; behind, the left common iliac vein, the right common iliac vein, and the beginning of the inferior vena cava; to the right, the right common

iliac vein and the psoas magnus muscle. The left common iliac
artery is in relation with the following structures: in front, the
peritoneum, the ureter, the inferior mesenteric artery, the sigmoid
flexure of the colon, and the ovarian artery in the female;
behind, the fourth and the fifth lumbar vertebræ and part of
the psoas magnus muscle; to the left, the psoas magnus muscle; to the right, the left common iliac vein. (Morris, p. 573;
Gray, p. 608.)

THE EXTERNAL ILIAC ARTERY.

The **external iliac artery** is a branch of the common
iliac artery at the sacro-iliac articulation. It passes forward, along
the inner border of the psoas magnus muscle, and beneath
Poupart's ligament, where it becomes the femoral artery.

RELATIONS.—In front, the peritoneum, the ileum on the
right, the sigmoid flexure of the colon on the left, the genito-
crural nerve, the vas deferens in the male, and the ovarian
vessels in the female. Behind, the psoas magnus muscle. Internally, the external iliac vein. Externally, the psoas magnus
muscle.

The **branches** of the external iliac artery are: (1) the *deep
epigastric* and (2) the *deep circumflex iliac.*

The **deep epigastric artery** passes upward and inward,
beneath the peritoneum. It lies above and to the outer side of
the femoral ring and to the inner side of the internal abdominal
ring. It enters the sheath of the rectus muscle at the semilunar
fold of Douglas and, in the substance of that muscle, it
terminates by anastomosing with the superior epigastric branch
of the internal mammary artery. Just after it is given off from
the external iliac artery, it is crossed by the vas deferens in
the male and by the round ligament in the female. It gives
off the **cremaster artery** to the cremaster muscle, muscular
branches to the abdominal muscles, and cutaneous branches to
the skin of the anterior abdominal wall.

The **deep circumflex iliac artery** passes outward, parallel
to the crest of the ilium, to terminate by anastomosing with

the fourth lumbar artery. In its course it lies, first, between the peritoneum and the transversalis fascia, and then between the transversalis and the internal oblique muscles. It gives branches to the muscles in its course and to the skin of the abdomen. (Morris, p. 600; Gray, p. 628.)

THE INTERNAL ILIAC ARTERY.

The **internal iliac artery** is a branch of the common iliac artery at the sacro-iliac articulation. It passes over the ilio-pectineal line and enters the true pelvis. It divides into an *anterior trunk* and a *posterior trunk* from which branches come off which supply the pelvic viscera, the pelvic outlet, and the gluteal region.

RELATIONS.—In front, the ureter and the peritoneum. Behind, the internal iliac vein, the lumbo-sacral cord, and the pyriformis muscle.

The **branches** of the internal iliac artery are: (a) from the posterior trunk, (1) the *ilio-lumbar*, (2) the *lateral sacral*, (3) the *gluteal;* (b) from the anterior trunk, (4) the *superior vesical*, (5) the *middle vesical*, (6) the *inferior vesical*, (7) the *middle hemorrhoidal*, (8) the *obturator*, (9) the *sciatic*, (10) the *internal pudic;* (c) in the female, (11) the *uterine*, and (12) the *vaginal*.

The **ilio-lumbar artery** supplies the iliacus, the psoas, and the quadratus lumborum muscles, the sacro-iliac joint, and the vertebræ.

The **lateral sacral artery** passes downward along the margin of the sacrum and anastomoses with the middle sacral artery.

The **gluteal artery** leaves the pelvis by passing through the great sacro-sciatic foramen, above the tendon of the pyriformis muscle, in company with the superior gluteal nerve. It supplies the tissues in the gluteal region.

The **superior vesical artery** passes forward to the fundus of the bladder and is distributed to that organ. In the fetus, this vessel continues from the fundus of the bladder, beneath the peritoneum, to the umbilicus as the **hypogastric artery**. In the adult, the portion of the vessel from the fundus of the bladder to the umbilicus becomes obliterated and remains beneath the peritoneum as a fibrous cord, forming the boundary between the middle and the internal inguinal fossæ (see p. 152).

The **middle** and the **inferior vesical arteries** supply the bladder.

The **middle hemorrhoidal artery** is distributed to the rectum.

The **obturator artery** leaves the pelvis by passing through the obturator foramen, in company with the obturator nerve. It sends branches to the ilium, the pubes, the bladder, the hip joint, the obturator and the adductor muscles.

The **sciatic artery** leaves the pelvis by passing through the great sacro-sciatic foramen, below the tendon of the pyriformis muscle, in company with the sciatic nerves. It passes down the back of the thigh and ends in the crucial anastomosis. In its course down the thigh it lies successively upon the gemellus superior, the obturator internus, the gemellus inferior, the quadratus femoris, and the adductor magnus muscles. It gives branches to the muscles with which it is in relation, to the skin over the coccyx, to the skin in the gluteal region, to the great sciatic nerve, and to the hip joint.

The **internal pudic artery** leaves the pelvis by passing through the great sacro-sciatic foramen; below the tendon of the pyriformis muscle, in company with the internal pudic nerve. It passes across the spine of the ischium and enters the ischio-rectal fossa, by passing through the lesser sacro-sciatic foramen. In the ischio-rectal fossa, it lies in Alcock's canal, close to the ischium. It pierces the superior layer of the triangular ligament and, in the deep perineal interspace, divides into its terminal branches.

The **branches** of the internal pudic artery are: (1) the *inferior hemorrhoidal*, (2) the *superficial perineal*, (3) the *artery of the bulb*, (4) the *artery of the corpus cavernosum*, and (5) the *dorsal artery of the penis*.

The **inferior hemorrhoidal artery** supplies the anus.

The **superficial perineal artery** supplies the tissues closing in the pelvic outlet.

The **artery of the bulb** pierces the inferior layer of the triangular ligament and supplies the corpus spongiosum of the penis.

The **artery of the corpus cavernosum** pierces the inferior layer of the triangular ligament and supplies the corpus cavernosum of the penis.

The **dorsal artery of the penis** pierces the inferior layer of the triangular ligament, passes through the suspensory ligament of the penis, and runs down along the dorsum of that organ to terminate in an anastomotic circle around the glans.

The **uterine artery** passes between the layers of the broad ligament to the uterus.

The **vaginal arteries,** three in number, supply the vagina. (Morris, p. 587; Gray, p. 620.)

THE GREAT VEINS.

The external iliac vein and the internal iliac vein unite to form the **common iliac vein.**

The right common iliac vein and the left common iliac vein unite to form the inferior vena cava. The **inferior vena cava** passes upward along the posterior abdominal wall, resting on the bodies of the lumbar vertebræ, to the right of the median line. It passes through a fissure on the posterior surface of the right lobe of the liver, through the quadrate opening in the diaphragm, and, piercing the pericardium immediately, empties into the right auricle.

RELATIONS.—In front, the peritoneum, the right spermatic artery, the transverse colon, the mesentery, the transverse portion of the duodenum, the head of the pancreas, the portal vein, and the liver. Behind, the lumbar vertebræ, the right crus of the diaphragm, and the right renal artery. To the right, the liver. To the left, the abdominal aorta.

In its course, the inferior vena cava receives the right spermatic vein, the renal veins, the hepatic veins, the phrenic veins, the suprarenal veins, and the lumbar veins.

The **left spermatic vein** empties into the left renal vein. (Morris, p. 656; Gray, p. 673.)

The internal jugular vein and the subclavian vein unite to form the innominate vein. The **right innominate vein** is the shorter of the two veins. It passes through the superior mediastinum to join with the left innominate vein.

RELATIONS.—In front, the thymus gland. Behind, the right pleura. To the right, the right phrenic nerve. To the left, the right pneumogastric nerve and the innominate artery.

The **left innominate vein** is longer than the right. It passes transversely across the superior mediastinum to join with the right innominate vein. For this reason, it was called by Leidy the great transverse vein.

RELATIONS.—In front, the thymus gland and the first piece of the sternum. Behind, the left subclavian artery, the left common carotid artery, the innominate artery, the left phrenic nerve, the left pneumogastric nerve, and the trachea. Below, the transverse portion of the arch of the aorta.

The innominate veins receive the vertebral veins, the inferior thyroid veins, and the internal mammary veins. The **left superior intercostal vein** empties into the left innominate vein. The **right superior intercostal vein** empties into the vena azygos major.

The two innominate veins unite in the superior mediastinum to form the **superior vena cava.** The superior vena cava receives the vena azygos major and empties into the right auricle.

RELATIONS.—In front, the thymus gland, the pleura, and the pericardium. Behind, the vena azygos major, the right bronchus, the right pulmonary artery, and the right superior pulmonary vein. To the right, the phrenic nerve. To the left, the innominate artery and the ascending portion of the arch of the aorta. (Morris, p. 627; Gray, p. 665.)

THE AZYGOS VEINS.

The **vena azygos major** begins at the confluence of the right lumbar veins, behind the right renal vein. It passes through the aortic opening in the diaphragm, upward through the posterior mediastinum, and, winding over the root of the right lung, empties into the posterior portion of the superior vena cava, just before that vessel pierces the pericardium. It receives the vena azygos minor, the right intercostal veins, the right superior intercostal vein, and the right bronchial vein.

The **vena azygos minor** begins at the confluence of the left lumbar veins, behind the left renal vein. It enters the posterior mediastinum by passing behind the left crus of the diaphragm. It passes across the body of the eighth thoracic vertebra to empty into the vena azygos major. It receives the four lower left intercostal veins and the left upper azygos vein.

The **left upper azygos vein** begins in the fifth or sixth left intercostal vein and passes downward in the posterior mediastinum, to empty into the vena azygos minor or into the vena azygos major. It receives the fifth, sixth and, seventh left intercostal veins. (Morris, p. 630; Gray, p. 667.)

THE PORTAL VEIN.

The **portal vein** is formed behind the pancreas by the union of the splenic vein and the superior mesenteric vein. It passes through the lesser omentum, lying between and behind the common bile duct and the hepatic artery. It enters the liver by passing through the transverse fissure of that organ. The **inferor mesenteric vein** empties into the splenic vein. The **gastric vein** empties into the portal vein. The blood from the tributaries of these four veins, the superior mesenteric the inferior mesenteric, the splenic, and the gastric, finally enters the liver. These veins and their tributaries form the portal system. (Morris, p. 659; Gray, p. 649.)

THE PULMONARY VEINS.

The **pulmonary veins** are four in number: the right superior, the right inferior, the left superior, and the left inferior. They empty into the left auricle of the heart. (Morris, p. 626; Gray, p. 650.)

THE DEVELOPMENT OF THE CARDIO-VASCULAR SYSTEM.

THE HEART.—**The heart** is developed in the mesoderm of the splanchnopleure. It makes its appearance, shortly after the celom is formed, as two tubes, one of which is situated on either side of the axis of the embryo. As the splanchnopleure folds anterioly to form the gut tract, the two tubes approach each other and finally fuse. As a result of this fusion, the first indication of a centrally placed heart is a straight, endothelial tube which lies just beneath the foregut. The muscular heart develops around the endothelial heart, the two being separated by a space which is crossed by trabeculæ of embryonal connective tissue. The inferior portion of the straight heart is venous, the superior portion is arterial. The heart rapidly increases in size and, in order to accommodate itself to its limited space, it becomes twisted both from behind forward and from above downward. In this process of twisting, which results in the formation of an S-shaped figure, the arterial portion of the heart grows downward and in front of the venous portion, which, in turn, grows upward and backward. As the venous portion of the heart grows upward, it gives off two processes which appear on either side of the arterial portion. These are the future **auricular appendages,** the first portions of the auricles to be formed. With the development of the lungs, septa grow from above downward and from below upward, dividing the cavity of the heart into a right portion and a left portion. The septum which grows from above downward, presents an opening which is not closed until about the time of birth, the **foramen ovale.** A lateral septum grows across the cavity of the heart, dividing the auricles from the ventricles. This septum presents an opening on each side of the heart, the **auriculo-ventricular opening.** The **valves** are formed by the growth of the endocardium in the neighborhood of these openings.

In the early stages, the veins unite outside the heart to form a common trunk, the **sinus venosus.** As the heart grows, this tube is included in the cavity of the right auricle.

The **truncus arteriosus** arises from the ventricular portion of the heart, in the early condition. When the heart is divided into a right and a left portion, the truncus arteriosus also becomes divided into an anterior vessel, the **pulmonary artery,** and a posterior vessel, the **aorta.** The truncus arteriosus was originally guarded by four folds of endocardium. When the division takes place, two of these valves are divided so that the pulmonary artery is provided with one posterior and two anterior semilunar valve leaflets and the aorta receives one anterior and two posterior semilunar valve leaflets.

The areolar tissue which separates the endothelial heart from the muscular heart guides the growth of muscular bands along its trabeculæ, producing the **columnæ carneæ** and the **musculi papillares.**

THE BLOOD AND THE BLOOD VESSELS.—The **blood vessels** are formed outside the body of the embryo and grow into the embryonic area, to join the venous portion of the developing heart. The mesodermic cells covering the umbilical vesicle become grouped into irregular areas which give rise to the blood and the blood vessels. These areas are known as the **blood islands of Pander.** The cells forming the periphery of these islands branch and form the walls of the vessels. Of the cells in the centre of these areas, some become liquefied to form the blood plasma, and others persist as the fetal red blood corpuscles, which are nucleated. The vessels thus formed unite to form two large trunks which are known as the **vitelline veins.** They join, just outside the heart, to form the sinus venosus. When the placental circulation is established, the **right** and the **left allantoic veins,** from the placenta, empty into the sinus venosus together with the **right** and the **left veins of Cuvier.** The veins of Cuvier are formed by the union of the anterior cardinal and the posterior cardinal veins. The vitelline veins become the **portal vein** and the **hepatic veins** in the adult. The right and left allantoic veins fuse to form the **umbilical vein.** This vessel carries the blood from the placenta to the fetus. After passing beneath the peritoneum, from the umbilicus to the under surface of the liver, it empties into the portal vein, as the latter vessel enters the liver. At a later period, a short cut, the **ductus venosus** is established between the portal vein and the inferior vena cava.

The **anterior cardinal vein** becomes the **external jugular vein,** in the adult. The **internal jugular vein** is formed as a new trunk. The **posterior cardinal vein** drains the Wolffian body of the embryo. When the kidney is formed, the posterior cardinal veins become smaller in proportion and remain in the adult as the **azygos veins,** major and minor. The **internal iliac** veins develop from the tips of the posterior cardinal veins.

The **inferior vena cava** is developed simultaneously with the kidneys, partly as a new vessel and partly by utilizing veins which already exist. The new portion of the vessel originates from the ductus venosus. In its growth, the inferior vena cava includes within itself a portion of the right posterior cardinal vein. The **right common iliac vein** represents a portion of the right posterior cardinal vein. The **external iliac veins** and the **left common iliac veins** are new vessels.

The **veins of Cuvier** are also known as the **superior venæ cavæ.** Originally, there are two of these vessels; but the left one soon atrophies, its end persisting in the adult as the **coronary sinus.** The right vein of Cuvier persists as the adult **superior vena cava.** When the left vein of Cuvier atrophies a transverse communication is established from the left side to the remaining superior vena cava. This trunk then becomes the **left innominate vein.**

THE ARTERIES.—Two vessels, the **ventral aortæ,** arise from the truncus arteriosus, and pass upward, along the ventral surface of the embryo, to the position of the first visceral arches of either side. The arterial trunks pass through these arches and then pass downward on either side of the vertebral column as the **dorsal aortæ.** The ventral and the dorsal aortæ are connected by vascular arches which pass through the five visceral arches already described (see p. 14).

The first aortic arch disappears. The ventral aorta helps

to form the **external carotid artery**, the dorsal aorta helps to form the **internal carotid artery.**

The second aortic arch disappears. The ventral aorta and the dorsal aorta enter into the formation of the external and the internal carotid arteries, as in the first arch.

The third aortic arch persists and completes the **internal carotid artery.** The dorsal aorta belonging to this arch disappears. The ventral aorta forms the **common carotid artery.**

The fourth aortic arch, on the right side, becomes the **subclavian artery;** its dorsal aorta disappears; its ventral aorta forms the **innominate artery.** On the left side, the fourth aortic arch forms the **arch of the aorta;** its dorsal aorta becomes the adult **thoracic aorta;** its ventral aorta forms the **ascending portion** of the arch of the aorta.

The fifth aortic arch, on the right side, forms the **right pulmonary artery;** its dorsal aorta disappears. On the left side, the fifth aortic arch becomes the **left pulmonary artery** and the **ductus arteriosus.** Below the fifth aortic arch the dorsal aortæ fuse to form the **thoracic** and **abdominal aortæ.** (Quain, p. 134; A. T. O., p. 103.)

THE FETAL CIRCULATION.

In the course of development and of independent existence, the human animal has three different circulations, the vitelline, the placental, allantoic, or fetal, and the adult. In the **fetal circulation** the blood is brought to the body of the fetus from the placenta by the umbilical vein. Some of this blood passes through the liver to the hepatic veins and is then emptied into the inferior vena cava; the remainder of the blood enters the inferior vena cava by passing through the ductus venosus. The blood, mixed with that returned from the lower extremities, then enters the right auricle and, guided by the Eustachian valve, passes into the left auricle through the foramen ovale. From the left auricle it passes successively through the left ventricle, the aorta, the hypogastric arteries, and the umbilical arteries to the placenta.

The blood is returned to the right auricle from the head and the upper extremities by the superior vena cava. It passes from the superior vena cava through the right auriculo-ventricular opening, guided by the tubercle of Lower, into the right ventricle. From the right ventricle it passes into the pulmonary artery. Some of this blood goes on to the lungs; but the larger portion of it enters the aorta, by passing through the ductus arteriosus, and then takes the course of the remainder of the blood in the aorta, to the placenta. The blood that reaches the lungs is returned to the left auricle by the pulmonary veins.

After birth the foramen ovale closes and becomes the fossa ovalis of the adult. The umbilical vein becomes obliterated and forms the round ligament of the liver. The ductus venosus and the ductus arteriosus become obliterated. The hypogastric artery becomes obliterated from the fundus of the bladder to the umbilicus. The patulous portion of the hypogastric artery is then known as the superior vesical artery. (Quain, p. 155; A. T. O., p. 136; Morris, p. 956; Gray, p. 1097.)

CHAPTER XIV.

THE RESPIRATORY SYSTEM.

The **respiratory system** consists of the following parts: the *nose*, the *larynx*, the *trachea*, the *bronchi*, and the *lungs*.

THE NOSE.

The **nose** may be divided into an *external* and an *internal* portion.

The **external portion of the nose** is formed by the nasal bones, the superior lateral cartilages, and the inferior lateral cartilages. This framework is covered by skin, connective tissue, and certain muscles.

The **internal portion of the nose** is spoken of as the **nasal fossæ.** The nasal fossæ are bounded by a roof, a floor, an outer wall, and an inner wall or septum. The roof is formed by the nasal bones, the cribriform plate of the ethmoid bone, and the inferior surface of the body of the sphenoid bone. The outer wall is formed by the lachrymal bone, the superior maxillary bone, the vertical plate of the palate bone, the os planum of the ethmoid bone, and the internal pterygoid plate of the sphenoid bone. The floor is formed by the palate process of the superior maxillary bone and the horizontal plate of the palate bone. The septum is formed by the perpendicular plate of the ethmoid bone, the vomer, and the cartilage of the septum.

The nasal fossæ open anteriorly by the **anterior nares**, which are bounded *above*, by the nasal bones; *laterally*, by the free edges of the superior maxillary bones; and *below*, by the anterior nasal spine. Posteriorly, they empty into the pharynx through the **posterior nares**. The posterior nares are bounded, *internally*, by the vomer; *externally*, by the internal pterygoid plate of the sphenoid bone; *above,* by the body of the sphenoid bone; and *below,* by the horizontal plate of the palate bone.

Each nasal fossa is divided into three meatuses by the turbinated bones. The **superior meatus** is situated between the superior and the middle turbinated bones. The **middle meatus** is situated between the middle and the inferior turbinated bones. The **inferior meatus** is situated between the inferior turbinated bone and the floor of the nose.

In the **superior meatus** of the nose we see the openings of the sphenoidal air cells and of the posterior ethmoidal cells.

In the **middle meatus** of the nose we see the openings of the frontal air cells, of the anterior ethmoidal cells, and of the antrum of Highmore. The frontal and the anterior ethmoidal cells open into the middle meatus by a common passageway, which is known as the **infundibulum.** The antrum of Highmore is also known as the **maxillary sinus**.

In the **inferior meatus** of the nose we see the opening of the nasal duct, which brings the tears from the conjunctiva.

The **olfactory fissure** is the narrow passageway between the superior turbinated bone and the septum of the nose.

The mucous membrane which lines the cavity of the nose may be divided into the **olfactory portion** and the **respiratory portion.** The olfactory portion of the nasal mucous membrane is confined entirely to the superior meatus of the nose.

The nasal fossæ are supplied by the following **nerves:** the olfactory, to the superior meatus; the nasal, to the under surface of the nasal bone and to the adjacent portions of the outer wall and the septum; the naso-palatine to the vomer and the adjacent parts; and the anterior palatine nerve and branches of the anterior superior dental nerve, to the turbinated bones.

The following **arteries** send twigs to the nose; the anterior ethmoidal, the posterior ethmoidal, the naso-palatine, and the descending palatine. (Morris, pp. 98 and 905; Gray, pp 219 and 885.)

THE LARYNX.

The **larynx** is a cartilaginous box which contains the structures concerned in the production of voice.

The cartilages composing the larynx are: (1) the *thyroid,* (2) the *cricoid,* (3) the *epiglottis,* (4 and 5) the *arytenoids,* (6 and 7) the *cartilages of Wrisberg,* and (8 and 9) the *cartilages of Santorini.*

The **thyroid cartilage** consists of two broad, lateral plates or alæ, which are joined, anteriorly, to form an acute angle which is known as **Adam's apple.** The superior border of this cartilage presents the **thyroid notch,** anteriorly, and the **superior cornu** at either end. From the inferior border of the thyroid cartilage, the **inferior cornua** spring.

The **cricoid cartilage** is shaped like a seal ring. Its narrower portion is situated anteriorly; its wider portion is placed posteriorly. It articulates with the inferior cornua of the thyroid cartilage, forming the crico-thyroid articulation.

The **epiglottis** is attached to the inner surface of the anterior portion of the thyroid cartilage. It projects upward, behind the base of the tongue and behind the body of the hyoid bone. The epiglottis is connected to the tongue by the three **glosso-epiglottic folds.**

The **arytenoid cartilages** are pyramidal in shape. They articulate by their bases with facets which are seen on the superior surface of the cricoid cartilage. The bases of the arytenoid cartilages present three angles. The true vocal cords are attached to the anterior angles or **vocal processes ;** the crico-arytenoideus muscles are inserted into the external angles ; the internal angles are of less importance. The apex of each of arytenoid cartilage is blunt.

The **cartilages of Wrisberg** are found in the aryteno-epiglottic folds.

The **cartilages of Santorini** are found surmounting the apices of the arytenoid cartilages.

The thyroid cartilage is attached to the hyoid bone by its superior cornua, and by the **thyro-hyoid membrane.**

The cricoid cartilage is connected to the thyroid cartilage by the **crico-thyroid membrane.** Laterally, this membrane projects into the larynx and is attached, anteriorly, to the thyroid cartilage; and posteriorly, to the anterior angle (vocal process) of the arytenoid cartilage. Between these two points of attachment, the crico-thyroid membrane presents a sharp, free edge which is covered by mucous membrane and which constitutes the true vocal cord.

The mucous membrane which lines the larynx is thrown into several well-marked folds. The **aryteno-epiglottic folds** extend from the tips of the arytenoid cartilages to the point of attachment of the epiglottis to the thyroid cartilage.

The **false vocal cords** are two folds of the laryngeal mucous membrane which pass from the anterior surfaces of the arytenoid cartilages to the thyroid cartilage. They are situated above the true vocal cords and do not project so far into the lumen of the larynx.

The **true vocal cords** extend from the anterior angles of the bases of the arytenoid cartilages to the thyroid cartilage. The basis of the true vocal cord is the free edge of the crico-thyroid membrane.

The **ventricle of the larynx** is the pouch between the false and the true vocal cords.

The **glottis** is the chink between the true vocal cords.

The **superior opening of the larynx** is bounded, *in front*, by the epiglottis; *behind*, by the tips of the arytenoid cartilages; and *laterally*, by the aryteno-epiglottic folds.

The **intrinsic muscles** of the larynx are: (1) the *crico-thyroid*, (2) the *posterior crico-arytenoid*, (3) the *lateral crico-arytenoid*, (4) the *thyro-arytenoid*, and (5) the *arytenoid*.

The larynx is supplied with **blood** by the superior laryngeal branch of the superior thyroid artery and by the inferior laryngeal branches of the inferior thyroid artery.

The **nerves** which are distributed to the larynx come from the pneumogastric. The superior laryngeal nerve supplies the mucous membrane of the larynx and gives the external laryngeal branch to the crico-thyroid muscle. The inferior or recurrent laryngeal nerve supplies all the muscles except the crico-thyroid and sends a branch to the mucous membrane. (Morris, p. 917; Gray, p. 1100.)

THE TRACHEA.

The **trachea** begins at the lower border of the sixth cervical vertebra and ends at the lower border of the fourth thoracic vertebra by dividing into two bronchi. In its course it passes through the superior mediastinum. It is about four inches in length

RELATIONS.—In the neck, the trachea is in relation, posteriorly, with the esophagus; on either side, with the lateral masses of the thyroid body, the recurrent laryngeal nerve, and the sheath of the carotid blood vessels; anteriorly, with the isthmus of the thyroid body. In the superior mediastinum, it is in relation, posteriorly, with the esophagus; to the right, with the innominate artery and the right pneumogastric nerve; to the left, with the left common carotid artery and the left pneumogastric nerve; anteriorly, with the left innominate vein, the arch of the aorta, and the remains of the thymus gland. The cartilaginous rings of which the trachea is composed are incomplete for the posterior one-third of the circumference of the tube. (Morris, p. 929; Gray, p. 1108.)

THE BRONCHI.

The **bronchi,** two in number, are given off from the trachea at the lower border of the fourth thoracic vertebra. They pass laterally to enter the root of the lung.

The **right bronchus** is shorter; but of larger diameter than is the left. It seems to be the direct continuation of the trachea. It is about one inch long. The right bronchus divides into three branches, one of which passes to each lobe of the right lung. The tube which goes to the superior lobe of the right lung passes above the pulmonary artery and is known as the **eparterial bronchus.**

RELATIONS.—In front, the superior vena cava, the right pulmonary artery, and the ascending portion of the arch of the aorta. Behind, the vena azygos major.

The **left bronchus** is about two inches in length. It passes beneath the arch of the aorta to enter the root of the left lung.

The left bronchus divides into two branches, one to each lobe of the left lung.

RELATIONS.—In front, the arch of the aorta and the left pulmonary artery. Behind, the esophagus, the thoracic duct, and the descending portion of the arch of the aorta. (Morris, p. 931; Gray, p. 1108.)

THE LUNGS.

Each lung is invested by a serous membrane which is called the **pleura.** The pleura presents a **parietal layer,** which lines the walls of the thoracic cavity, and a **visceral layer** which is reflected along the structures forming the root of the lung and which closely invests that organ except at the hilum. The inner surface of the pleura is in relation with the pericardium. It is in relation, below, with the diaphragm. The pleura is reflected from the thoracic wall (**costal pleura**) to the diaphragm (**diaphragmatic pleura**) in a line which is drawn from the seventh costal cartilage, anteriorly, obliquely across the eighth, ninth, tenth, and eleventh ribs to the posterior axillary line. Posteriorly, the pleural sac extends to the twelfth rib and is, therefore, in close relation to the superior extremity of the kidney.

The **right lung** presents three lobes: a superior lobe, a middle lobe, and an inferior lobe. The middle lobe is a subdivision of the superior lobe; although, morphologically, it probably represents the superior lobe of the left lung, the superior lobe of the right lung being an additional structure. The **apex of the lung** extends for about an inch into the subclavian triangle in the neck. It presents a groove which is made by the subclavian artery. The lower border of the right lung extends as far down, posteriorly, as the lower border of the tenth rib. Anteriorly, the right lung ends at the sixth rib.

The **left lung** presents two lobes: a superior lobe and an inferior lobe. The **apex** of the left lung extends for about an inch into the subclavian triangle and presents a groove formed by the subclavian artery. The lower border of the left lung corresponds to the seventh rib, anteriorly, and to the eleventh rib, posteriorly.

The **root of the lung** is formed by the pulmonary vein, the pulmonary artery, and the bronchus. From before backward these structures are found in the following order on both sides: vein, artery, bronchus. From above downward the arrangement of the structures constituting the root of the lung differs on the two sides, on account of the longer course taken by the left bronchus in passing beneath the arch of the aorta. On the right side, we find bronchus, artery, vein; on the left side, artery, bronchus, vein.

The surface of each lung is divided into hexagonal areas, indicating the lobules into which the organ is divided. The right lung weighs about twenty-two ounces (660 grams); the left lung weighs about twenty ounces (600 grams).

The lungs are supplied with **nutriment** by the bronchial arteries (see p. 190).

The **nerves** of the lung are derived from the anterior and the posterior pulmonary plexuses (see p. 74). (Morris, p. 936; Gray, p. 1113.)

THE MEDIASTINAL SPACES.

That portion of the thorax which is not occupied by the pleuræ and the lungs is spoken of as the **mediastinum**. This space is subdivided into the *superior mediastinum* and the *inferior mediastinum*. The inferior mediastinum is again subdivided into the *anterior mediastinum,* the *middle mediastinum,* and the *posterior mediastinum.*

The **superior mediastinum** is bounded, *above*, by a plane passed from the superior margin of the first piece of the sternum to the upper border of the first thoracic vertebra; *below*, by a plane passed from the point of union of the first and the second pieces of the sternum to the lower border of the fourth thoracic vertebra; *anteriorly*, by the first piece of the sternum; *posteriorly*, by the anterior surfaces of the first four thoracic vertebræ; and *laterally*, by the pleuræ. It contains the *arch of the aorta,* the *innominate artery,* the *left common carotid artery,* the *left subclavian artery,* the *right* and the *left innominate veins,* the *superior vena cava,* the *phrenic nerves,* the *pneumogastric* *nerves*, the *left recurrent laryngeal nerve*, the *cardiac nerves*, the *esophagus*, the *trachea*, the *thoracic duct*, and the remains of the *thymus gland*.

The **inferior mediastinum** is bounded, *above*, by a plane passed from the point of union of the first and second pieces of the sternum to the lower border of the fourth thoracic vertebra ; *below*, by the diaphragm ; *laterally*, by the pleuræ ; *anteriorly*, by the second piece of the sternum ; and *posteriorly*, by the anterior surfaces of the bodies of the thoracic vertebræ, from the fifth to the twelfth, inclusive.

The **anterior mediastinum** is that portion of the inferior mediastinum which lies in front of the anterior surface of the pericardium. It contains the *left internal mammary artery*, the *triangularis sterni muscle*, and *lymphatics*.

The **middle mediastinum** is that portion of the inferior mediastinum which is included between the anterior and the posterior surfaces of the pericardium. It contains the *heart*, the *ascending portion of the arch of the aorta*, the *pulmonary artery*, the *superior vena cava*, the *vena azygos major*, the *bifurcation of the trachea*, the *phrenic nerves*, and the *pulmonary veins*.

The **posterior mediastinum** is that portion of the inferior mediastinum which is situated behind the posterior surface of the pericardium. It contains the *thoracic aorta*, the *beginnings of the intercostal arteries*, the *vena azygos major*, the *vena azygos minor*, the *left upper azygos vein*, the *esophagus*, the *thoracic duct*, and the *pneumogastric nerves*. (Morris, p. 915 ; Gray, p. 1114.)

THE DIAPHRAGM.

The **diaphragm** is the broad sheet of muscular tissue which separates the thorax from the abdomen. It **arises from** the ensiform process of the sternum, from the lower borders of the cartilages of the seventh, eighth, ninth, tenth, eleventh, and twelfth ribs, from the ligamentum arcuatum externum, from the ligamentum arcuatum internum, and from the crura. It is **inserted into** a central tendon. It is **supplied by** the right and the left phrenic nerves.

The **ligamentum arcuatum externum** extends from the tip of the twelfth rib to the transverse process of the second lumbar vertebra, arching over the quadratus lumborum muscle.

The **ligamentum arcuatum internum** extends from the transverse process to the body of the second lumbar vertebra, arching over the psoas magnus muscle.

The **crura** of the diaphragm are two muscular masses which are attached to the bodies of the first, second, and third lumbar vertebræ, on the right, and to the bodies of the first and second lumbar vertebræ on the left.

The diaphragm presents three **openings,** the **quadrate,** which transmits the inferior vena cava, is the most anterior; the **esophageal,** which transmits the esophagus and the pneumogastric nerves, is placed in the middle; and the **aortic,** which transmits the aorta, the vena azygos major, and the thoracic duct, is the most posterior.

Behind the right crus, the greater and the lesser splanchnic nerves enter the abdomen; behind the left crus, the greater and the lesser splanchnic nerves and the vena azygos minor pass.

The phrenic nerves and the least splanchnic nerves pierce the diaphragm.

The superior epigastric atery passes between the diaphragm, the ensiform process of the sternum, and the seventh costal cartilage.

The diaphragm is in relation with the two pleuræ, the pericardium, and the peritoneum. (Morris, p. 419; Gray, p. 444.)

THE THYROID BODY.

The **thyroid body** is composed of two lateral lobes which are connected by an isthmus. This isthmus crosses the trachea at its second and third rings. The lateral masses are in relation with the trachea, the carotid blood vessels, and the recurrent laryngeal nerves. On the left side, the lateral mass is in relation with the esophagus. The thyroid body is intimately attached to the trachea and to the larynx and, consequently, moves with those organs. The thyroid body weighs about one ounce (30 grams). It is supplied with **blood** by the superior thyroid artery, a branch of the external carotid artery, and by the inferior thyroid artery, a branch of the thyroid axis. The **nerves** which pass to the thyroid body are derived from the middle cervical ganglion of the sympathetic nerve. (Morris, p. 931; Gray, p. 1122.)

THE THYMUS BODY.

The **thymus body,** in the adult, is a small, atrophic structure which is found in the superior mediastinum. It reaches maturity at the second year of extra-uterine life and subsequently undergoes retrograde changes. In the infant, it is composed of two lateral masses which extend from the lower border of the thyroid body to the fourth costal cartilage. When at its highest development, the thymus body weighs a little more than one dram (5 grams). (Morris, p. 935; Gray, p. 1124.)

THE DEVELOPMENT OF THE RESPIRATORY SYSTEM.

The **larynx,** the **trachea,** the **bronchi,** and the **lungs** are developed as outgrowths from the primitive gut. The **lungs** develop from the extremities of the branched bronchial tubes, much the same as is a racemose gland.

The **thymus body** develops from the third pharyngeal pouch (see p. 14).

The **thyroid body** develops from the fourth pharyngeal pouch (see p. 14) and from a downgrowth from the floor of the mouth. This body originally possessed a distinct duct which opened on the dorsum of the tongue. The remains of this duct are seen in the adult as the **foramen cecum** of the tongue. The middle lobe of the thyroid body is the true thyroid. (Quain p. 109; A. T. O., p. 118.)

CHAPTER XV.

THE DIGESTIVE SYSTEM.

The **digestive system** is composed of the *mouth*, the *pharynx*, the *esophagus*, the *stomach*, the *small intestine*, the *large intestine*, and the *glands* which empty their secretions into this tract.

THE MOUTH.

The **mouth** is bounded, *in front*, by the lips; *laterally*, by the cheeks; *above*, by the hard palate and the soft palate; and *below*, by the mylo-hyoid muscle.

The **lips** are composed of (1) the *skin*, (2) the *superficial fascia*, (3) the *orbicularis oris muscle*, and (4) the *mucous membrane*. The coronary arteries, superior and inferior, are found passing in the upper and lower lips, respectively, between the orbicularis oris muscle and the mucous membrane.

The **cheeks** are composed of (1) the *skin*, (2) the *superficial fascia*, (3) the *buccinator muscle*, and (4) the *mucous membrane*.

The **hard palate** is composed of (1) the *palate processes of the superior maxillary bones*, and (2) the *horizontal plates of the palate bones*.

The **soft palate** is composed of (1) the *tensor palati muscle*, (2) the *levator palati muscle*, (3) the *palato-glossus muscle*, (4) the *palato-pharyngeus muscle*, (5) the *azygos uvulæ muscle*, (6) the *mucous membrane of the mouth*, and (7) the *mucous membrane of the nose*.

The teeth and the alveolar arches, covered by the gums, divide the cavity of the mouth into the *vestibule* and the *mouth proper*. The **vestibule** is that portion of the mouth which is situated between the lips, in front, and the teeth and the gums, behind.

The **gums or gingivæ** are composed of mucous membrane and thickened periosteum.

The lips are attached to the gums by the **superior** and the **inferior frena,** folds of mucous membrane.

The **labial glands** are racemose glands which are seen in the mucous membrane of the lips.

The **buccal glands** are racemose glands which are situated in the mucous membrane of the cheeks. (Morris, p. 958; Gray, p. 930.)

The **tongue** is a muscular organ which is contained in the mouth. It is composed of a body, an anterior extremity or **tip,** and a posterior extremity or **root.** The ventral surface of the tongue is attached to the floor of the mouth by a fold of mucous membrane, known as the **frenum of the tongue.** The **muscles** which enter into the formation of the tongue are: (1) the *hyo-glossus,* (2) the *genio-hyo-glossus,* (3) the *palato-glossus,* (4) the *stylo-glossus,* and (5) the *lingualis.*

The tongue is covered by mucous membrane which, on the dorsal surface, presents numerous **papillæ** caused by projections of the tunica propria. The **conical** or **filiform papillæ,** the most numerous, are sharp and conical in shape. The **fungiform papillæ** resemble a truncated cone in appearance. They are less numerous than the filiform papillæ. The **circum-vallate papillæ,** from eight to twelve in number, are like large fungiform papillæ in appearance, each projecting papilla being surrounded by a ridge of mucous membrane and separated from this ridge by a shallow depression. The **taste buds** are found along the sides of the circumvallate papillæ. These papillæ are arranged in the form of a V, the apex of which is directed backward.

Just behind the apex of the figure formed by the circumvallate papillæ a small opening, the **foramen cecum,** may be seen. This is the remains of the thyro-glossal duct of the fetus (see p. 216).

That portion of the tongue which is situated behind the circumvallate papillæ contains a large amount of lymphoid tissue. This tissue is often spoken of as the **lingual tonsil.**

The following **arteries** supply the tongue: the dorsalis linguæ, to the dorsal aspect of the organ, particularly about the circumvallate papillæ; and the ranine artery, to the ventral surface of the tongue, as far forward as the tip of the organ. These vessels are branches of the lingual artery.

The **nerves** which supply the tongue are derived from three different sources. The hypoglossal nerve is the motor nerve, the glosso-pharyngeal nerve is the nerve of special sense, and the lingual branch of the inferior maxillary division of the tri-facial nerve is the nerve of common sensation. The chorda tympani nerve, which joins the lingual nerve, is probably composed of fibres of special sense, since it can be traced backward through the facial nerve and the pars intermedia to the deep origin of the glosso-pharyngeal nerve. (Morris, p. 900; Gray, p. 879.)

The **teeth** in the human animal make their appearance in two distinct sets. The first or *temporary teeth* and the second or *permanent teeth.*

A tooth is composed of a **root** or **fang,** which is received into a socket in the alveolar process of one of the maxillary bones, a **crown,** which projects beyond the gums, and a **neck,** which joins the crown to the fang. The tooth is held in the alveolar socket by a reflection of the periosteum from the bone to the root of the tooth.

The **temporary teeth** are twenty in number. In each half jaw we find a central incisor, a lateral incisor, a canine, and two molars.

There are thirty-two in the **permanent set.** In each half jaw we find a central incisor, a lateral incisor, a canine, two bicuspids, and three molars. The molar teeth are frequently spoken of as the **sixth-year molar, the twelfth-year molar,** and the **wisdom tooth,** respectively.

The upper teeth receive their **blood supply** from the posterior superior dental arteries, branches of the alveolar artery, which supply the molar and the bicuspid teeth; and from the anterior superior dental arteries, branches of the infraorbital artery, which supply the canine and the incisor teeth.

The **nerves** to the upper teeth are the posterior, middle, and anterior dental nerves, branches of the superior maxillary division of the trifacial nerve.

The lower teeth receive their **nutrition** from the inferior dental artery, a branch of the internal maxillary artery.

The **nerves** to the lower teeth are branches of the inferior dental branch of the inferior maxillary division of the trifacial nerve. A special nerve and a special artery, the incisive nerve and artery, pass to the lower incisor teeth. (Morris, p. 105; Gray, p. 932.)

THE ERUPTION OF THE TEMPORARY TEETH.

Central Incisor, 6 to 8 months.

Lateral Incisor, 7 to 10 months.

First Molar, 11 to 14 months.

Canine, 14 to 20 months.

Second Molar, 17 to 36 months.

THE ERUPTION OF THE ~~TEMPORARY~~ *Permanent* TEETH.

First Molar, 6th year.

Central Incisor, 7th year.

Lateral Incisor, 8th year.

First Bicuspid, 9th year.

Second Bicuspid, 10th year.

Canine, 11th year.

Second Molar, 12th year.

Third Molar, 18th to 25th year.

The **salivary glands** empty their secretions into the mouth. There are three pairs of these glands which are named, the *parotid,* the *submaxillary,* and the *sublingual.*

The **parotid gland** is a racemose gland. It is situated in front of the ear, below the zygoma, and superficial to the masseter muscle. A process of the parotid gland is found in the zygomatic fossa, between the external and the internal pterygoid muscles, the **pterygoid lobe;** a second process lies behind the articulation of the lower jaw, the **glenoid lobe;** and a third process is in relation with the carotid blood vessels, the **carotid lobe.** The parotid gland is separated from the sub-maxillary gland by the stylo-maxillary ligament.

The duct of the parotid gland, **Stenson's duct,** appears at the anterior margin of the gland. It passes transversely across the face, about one-half inch below the zygoma, in company with the transverse facial artery and the infraorbital branch of the facial nerve. It pierces the buccinator muscle and empties into the mouth opposite to the second upper molar tooth.

Passing through the substance of the parotid gland we find (1) the external carotid artery, (2) the posterior auricular artery, (3) the superficial temporal artery, (4) the internal maxillary artery, (5) the transverse facial artery, (6) the middle temporal artery, (7) the temporo-maxillary vein, (8) the posterior auricular vein, and (9) the facial nerve.

Beneath the parotid gland we find (1) the auriculo-temporal nerve, (2) the glosso-pharyngeal nerve, (3) the pneumogastric nerve, (4) the internal carotid artery, and (5) the internal jugular vein.

The **socia parotidis** is a small, isolated mass of parotid gland tissue which is found in front of the masseter muscle, about one-half inch from the anterior margin of the parotid gland. It empties its secretion into the duct of Stenson.

The **submaxillary gland** is a racemose gland which is situated beneath the angle of the inferior maxillary bone. It is separated from the parotid gland by the stylo-maxillary ligament and from the sublingual gland by the mylo-hyoid muscle. The

facial artery passes through a groove on the inferior surface of
the submaxillary gland and the facial vein lies on its superior
surface.

The duct of the submaxillary gland, **Wharton's duct,**
passes, in company with the lingual nerve, above the hyo-
glossus muscle. It then passes between the mylo-hyoid and
the genio-hyo-glossus muscles to empty on the summit of a
papilla which lies by the side of the frenum of the tongue.

The **sublingual gland** is a racemose gland which is situated
in the sublingual fossa of the inferior maxillary bone, just
beneath the mucous membrane of the mouth. It is separated
from the submaxillary gland by the mylo-hyoid muscle.

The large duct of the sublingual gland, the **duct of Bartholin,** empties in common with Wharton's duct. The smaller
ducts of the sublingual gland, fifteen or twenty in number, are
known as the **ducts of Rivini.** They empty into the floor of
the mouth. (Morris, p. 961 ; Gray, p. 945.)

The passageway from the mouth into the pharynx is known
as the **fauces.** The fauces is bounded on either side by the
anterior pillar and the *posterior pillar.* The **anterior pillar** of
the fauces is formed by the palato-glossus muscle; the **posterior pillar** of the fauces is formed by the palato-pharyngeus
muscle.

The **tonsil** is situated between the anterior and the posterior pillars of the fauces. The tonsil is a compound lymph
gland. The inner surface of the tonsil is in relation with the
fauces. Externally, the tonsil is separated from the internal
carotid and the ascending pharyngeal arteries by the superior
constrictor muscle of the pharynx. Anteriorly, the tonsil is in
relation with the palato-glossus muscle. Posteriorly, the tonsil
is in relation with the palato-pharyngeus muscle.

The tonsil is supplied by the following **arteries :** (1) the
tonsillar branch of the facial, (2) the dorsalis linguæ from the
lingual, (3) the ascending pharyngeal, a branch of the external
carotid, (4) the descending palatine branch of the internal maxillary, and (5) the ascending palatine branch of the facial.

The tonsil receives its **nerves** from the glosso-pharyngeal and from Meckel's ganglion. (Morris, p. 959; Gray, p. 945.)

THE PHARYNX.

The **pharynx** is a musculo-membranous bag which is suspended from the basilar process of the occipital bone and from the apices of the petrous portions of the temporal bones. It ends at the lower border of the fifth cervical vertebra. The soft palate divides the pharynx into the **naso-pharynx,** above that membrane, and the **oro-pharynx,** below that membrane.

There are seven **openings** into the pharynx: (1 and 2) the *posterior nares,* (3 and 4) the *Eustachian tubes,* (5) the *mouth,* (6) the *larynx,* and (7) the *esophagus.*

There is a collection of lymphoid tissue on the posterior wall of the pharynx, between the orifices of the Eustachian tubes, which is known as the **pharyngeal tonsil** or **Luschka's tonsil.**

The pharynx is formed by the *superior,* the *middle,* and the *inferior constrictor muscles,* the *palato-pharyngeus muscle,* and the *stylo-pharyngeus muscle.* The muscular coat of the pharynx is deficient in the upper portion of the organ between the apices of the petrous portions of the temporal bones and the basilar process of the occipital bone. The space is filled, however, by a strong layer of fibrous tissue which is known as the **pharyngeal aponeurosis.** The **raphé** of the pharynx is the median ridge formed by the interlacing of the constrictor muscles. It is situated on the posterior aspect of the organ and is attached, above, to the pharyngeal spine.

The following **arteries** supply the pharynx: the ascending pharyngeal branch of the external carotid, the pterygo-palatine branch of the internal maxillary, the descending palatine branch of the internal maxillary, the ascending palatine and the tonsillar branches of the facial, and the dorsalis linguæ branch of the lingual.

The **nerves** which supply the pharynx unite, on the middle constrictor muscle of the pharynx, to form the pharyngeal plexus. This plexus is formed by branches from the glosso-pharyngeal, the pneumogastric, and the spinal accessory nerves (see pp. 71, 73, and 81). (Morris, p. 964; Gray, p. 951.)

THE ESOPHAGUS.

The **esophagus** is a musculo-membranous tube which begins at the lower extremity of the pharynx, at the lower border of the fifth cervical vertebra, and ends at the cardiac extremity of the stomach, into which it empties. The esophagus is about nine inches long. In its course, the esophagus passes through the neck, lying to the left of the median line, through the superior mediastinum, lying in the median line, and through the posterior mediastinum, lying to the left of the median line. It passes through the esophageal opening in the diaphragm and terminates by emptying into the stomach. The esophagus presents constrictions opposite the sixth cervical vertebra and as it passes through the diaphragm.

RELATIONS.—In the neck, the esophagus is in relation, in front, with the trachea; behind, with the bodies of the sixth and seventh cervical vertebræ; on either side, with the recurrent laryngeal nerve, the lateral mass of the thyroid body, and the sheath of the carotid blood vessels. In the superior mediastinum, the esophagus is in relation, in front, with the trachea; behind, with the bodies of the first four thoracic vertebræ; to the right, with the pleura; to the left, with the pleura and the thoracic duct. In the posterior mediastinum, the esophagus is in relation, in front, with the posterior surface of the pericardium and the left pneumogastric nerve; behind, with the bodies of the thoracic vertebræ from the fifth to the tenth, inclusive, the thoracic duct, the thoracic aorta, and the left pneumogastric nerve; on the right, the vena azygos major and the pleura; on the left, the pleura and the thoracic aorta.

THE ABDOMINAL CAVITY.

The remainder of the digestive system, the stomach, the small intestine, and the large intestine, is contained in the abdominal cavity. This cavity is divided into nine **regions** by four lines. A perpendicular line is erected from the middle of Poupart's ligament on each side of the body. Transverse lines are drawn between the anterior superior spinous processes of the ilia and between the lower borders of the tenth costal cartilages. The spaces between these lines are given certain names. In the centre, we have, from above downward, the **epigastric region, the umbilical region,** and the **hypogastric region.** On either side, we have, from above downward, the **hypochondriac region, the lumbar region,** and the **inguinal** or **iliac region.**

The abdomen is lined by an extensive serous membrane which is termed the **peritoneum.** The peritoneum is reflected over all the viscera contained in the abdominal cavity, furnishing them with their means of attachment to the abdominal walls. On account of the transverse position of the stomach and of the transverse colon in the abdomen we have two folds of peritoneum, one in front of these organs, the **greater fold of peritoneum;** and one behind the stomach, the **lesser fold of peritoneum.** The **greater cavity** of the peritoneum is found between the visceral and the parietal layers of the greater fold of peritoneum. The **lesser cavity** of the peritoneum is found between the visceral and the parietal layers of the lesser fold of peritoneum. The **visceral layer** of the peritoneum is that portion of the membrane which encloses the viscera. The **parietal layer** of the peritoneum is that portion of the membrane which lines the abdominal wall.

In order to trace the **greater fold of peritoneum** we begin at the diaphragm. The peritoneum passes from the posterior portion of the diaphragm to the superior border of the liver, forming the anterior layer of the coronary ligament of the liver. It then passes over the superior surface of the liver to the anterior margin of that organ and over the inferior surface of the liver to its transverse fissure, where it is attached. From the transverse fissure of the liver the peritoneum passes forward to the lesser curvature of the stomach, forming the anterior layer of the gastro-hepatic omentum. It then passes in front of the stomach to the greater curvature of that organ, where it is attached. From the greater curvature of the stomach the peritoneum passes in front of the transverse colon to a free margin in front of the small intestine. It is then reflected upon itself and passes upward to the posterior surface of the transverse colon, from which surface it passes back to the posterior abdominal wall, where it is attached, forming the inferior layer of the transverse mesocolon. It then passes downward along the posterior abdominal wall until it reaches the superior mesenteric artery; it then passes over this artery and its branches, to the small intestine, encloses the small intestine in a reflection, and passes back to the posterior abdominal wall, completing the mesentery. It then passes downward along the posterior abdominal wall, over the promontory of the sacrum and down the posterior wall of the pelvis, enclosing the rectum. From the rectum, it passes across the floor of the pelvis to the posterior wall of the vagina, which it covers, and then, passing over the posterior surface, the fundus, and a portion of the anterior surface of the uterus, it passes forward to the posterior surface of the bladder. It covers the posterior surface of the bladder and then passes along the anterior abdominal wall to the diaphragm, over which it is reflected, to the point from which it started.

The pouch of peritoneum between the anterior surface of the rectum and the posterior wall of the vagina and the posterior wall of the uterus is known as the **pouch of Douglas**. The pouch of peritoneum between the anterior wall of the uterus and the posterior wall of the bladder is known as the **utero-vesical pouch**. In the male, the peritoneum is reflected from the antrior wall of the rectum, across the floor of the pelvis, to the posterior surface of the bladder, forming the **recto-vesical pouch**.

In order to trace the **lesser fold of the peritoneum** we begin at the diaphragm, behind the greater fold which we have already traced. From the diaphragm, the lesser fold of peritoneum passes to the transverse fissure of the liver directly, leaving the posterior surface of that organ nearly entirely devoid of a pentoneal coat, and forming the posterior layer of the coronary ligament of the liver. From the transverse fissure of the liver, it passes to the lesser curvature of the stomach, forming the posterior layer of the gastro-hepatic omentum. From the lesser curvature of the stomach, it passes over the posterior wall of that organ to the greater curvature of the stomach, where it is attached. From the greater curvature of the stomach, it passes in front of the transverse colon; but behind the layer of the greater fold of peritoneum already traced, to a free margin in front of the small intestine. It is then reflected on itself and again passes in front of the transverse colon, beneath the two layers already described. From the upper border of the transverse colon, it passes backward to the posterior abdominal wall where it is attached, forming the superior layer of the transvere mesocolon. It then passes upward along the posterior abdominal wall, in front of the transverse portion of the duodenum, the pancreas, and the great blood vessels, and over the crura of the diaphragm, to the place from which it started.

In tracing the peritoneum from side to side, we find a decided difference in the arrangement of the membrane in the upper and in the lower portion of the abdominal cavity.

If we trace the peritoneum from side to side at about the first lumbar vertebra, we expose both the greater and the lesser cavities of the peritoneum. Starting from the midline, anteriorly, the peritoneum passes over the anterior abdominal wall, to the right, until it reaches the round ligament of the liver, which it encloses in a reflection. It then passes around the lateral abdominal wall to the posterior abdominal wall, over which it passes in front of the kidneys, the pancreas, the transverse portion of the duodenum, and the great blood vessels. Opposite the fundus of the stomach it is reflected from the posterior abdominal wall to the posterior surface of that organ, which it covers. It passes around the pyloric end of the stomach, after sending a fold around the hepatic artery, the portal vein, and the common bile duct, and invests the anterior surface of that organ. From the fundus of the stomach it passes to the hilum of the spleen, forming the gastro-splenic omentum. It then invests the spleen and again reaching the hilum of that organ, this time at its posterior aspect, passes to the posterior abdominal wall. It then passes over the posterior, lateral, and anterior abdominal walls to the midline.

If we trace the peritoneum from side to side, just below the umbilicus, we find that, starting from the midline of the anterior abdominal wall, it covers the anterior and lateral abdominal walls, and the posterior abdominal wall until it reaches the position of the ascending colon. It encloses this portion of the gut in a fold, forming the ascending mesocolon. It then continues across the posterior abdominal wall, covering the abdominal aorta and the inferior vena cava and, sending a fold forward to include the small intestine, forms the mesentery. It then continues across the posterior abdominal wall until it reaches the descending colon, over which it is reflected, forming the descending mesocolon. It then passes over the lateral and anterior abdominal walls to the midline.

The lesser peritoneal cavity is situated between the posterior abdominal wall and the posterior wall of the stomach. It communicates with the greater peritoneal cavity through the **foramen of Winslow.** The foramen of Winslow is bounded, *above*, by the caudate lobe of the liver; *below*, by the hepatic artery; *anteriorly*, by the gastro-hepatic omentum; and *posteriorly*, by the inferior vena cava.

Folds of peritoneum connecting certain of the abdominal viscera to the abdominal walls and to each other receive special names. Thus, we speak of *omenta*, of *mesenteries*, and of *ligaments*.

An **omentum** is a double fold of peritoneum which connects the stomach with some adjacent organ. The omenta are three in number: the *greater omentum*, the *lesser omentum*, and the *least omentum.*

The **greater** or **gastro-colic omentum** extends from the greater curvature of the stomach to the transverse colon and thence to a free margin; its apron-like, free portion covers the small intestine. The greater omentum is composed of four layers of peritoneum; two layers from the greater fold and two layers from the lesser fold of peritoneum. Three of these layers, both layers of the lesser fold of peritoneum and one of the layers of the greater fold of peritoneum, lie in front of the transverse colon. One of these layers lies behind the transverse colon and helps to form the transverse mesocolon. Between the layers of the gastro-colic omentum we find the gastro-epiploica dextra and the gastro-epiploica sinistra arteries and their branches.

The **lesser** or **gastro-hepatic omentum** passes from the transverse fissure of the liver to the lesser curvature of the stomach. It is formed, in front, by a layer of the greater fold of peritoneum; and behind, by a layer of the lesser fold of peritoneum. It contains the hepatic artery, the portal vein, and the common bile duct. The artery is placed to the left, the bile duct to the right, and the portal vein is between and behind the other vessels.

The **least** or **gastro-splenic** omentum passes from the fundus of the stomach to the hilum of the spleen. It contains, between its layers, the vasa brevia, branches of the splenic artery.

A **mesentery** is a double fold of peritoneum which attaches some portion of the intestine to the posterior abdominal wall.

The **mesentery** proper is that fold of peritoneum which anchors the small intestine to the posterior abdominal wall. It is attached from the body of the second lumbar vertebra to the right sacro-iliac articulation, a distance of about six inches. At its intestinal border, the mesentery is about twenty feet in length. It contains the superior mesenteric artery and vein, the superior mesenteric plexus of sympathetic nerves, and the mesenteric lymphatics.

The **mesocolon** is the fold of peritoneum which attaches the colon to the posterior abdominal wall. The **ascending mesocolon** attaches the ascending colon to the abdominal wall. Between its layers we find the branches of the right colic and the ileo-colic arteries. The **transverse mesocolon** anchors the transverse colon to the posterior abdominal wall. It is composed of a layer of the greater fold of peritoneum, below, and a layer of the lesser fold of peritoneum, above. Between the layers of the transverse mesocolon the middle colic artery anastomoses with the right colic and the left colic arteries. The **descending mesocolon** binds the descending colon to the posterior abdominal wall. The left colic artery passes between its layers. The **mesosigmoid** anchors the sigmoid flexure of the colon to the posterior abdominal wall. It contains the sigmoid artery.

The **mesorectum** anchors the rectum to the posterior wall of the pelvis. It contains the superior and the middle hemorrhoidal arteries between its layers.

The **mesoappendix** is a fold of mesentery which carries the blood supply to the vermiform appendix. It is usually derived from the mesentery proper.

The cecum, as a rule, has no mesentery; if there is a mesentery provided for this part of the bowel it is known as the **mesocecum.**

The **duodeno-jejunal fossa** is produced by a fold of peritoneum which passes from the fourth portion of the duodenum to the jejunum.

The **superior ileo-cecal fossa** is formed by a fold of peritoneum which passes from the ileum to the cecum. It contains a branch of the ileo-colic artery.

The **inferior ileo-cecal fossa** is formed by a fold of peritoneum which passes from the under surface of the ileum, across the ileo-cecal region, to be attached to the mesoappendix. In order to expose it, the appendix should be pulled downward and the cecum should be pulled upward. (Morris, pp. 968 and 1223; Gray, p. 959.)

THE STOMACH.

The **stomach** is situated in the left hypochondriac and the epigastric regions. The **cardiac end** of the stomach is situated at the left extremity of the organ; the **pyloric end** of the stomach is placed to the right. The esophagus opens into the cardiac end of the stomach, while the pyloric end opens into the duodenum. The **fundus** of the stomach is all that portion of the cardiac end of the stomach to the left of the esophageal opening. The stomach presents a superior, **lesser curvature** and an inferior, **greater curvature**. The **pyloric antrum** is a dilatation in the lumen of the pyloric end of the stomach, seen just before the stomach opens into the duodenum. The pyloric opening of the stomach is guarded by a circular layer of muscle tissue, which is known as the **pyloric valve**. The mucous membrane of the stomach is thrown into folds or rugæ. The stomach is placed obliquely in the abdominal cavity and will, normally, hold about four pints.

RELATIONS.—Above, the stomach is in relation with the left lobe of the liver. Behind, the stomach is in relation with the left kidney, the left suprarenal body, the pancreas, the solar plexus, the abdominal aorta, and the inferior vena cava. In front, the stomach is in relation with the anterior abdominal wall in a triangular area which is bounded, *on the right*, by the lower edge of the liver; *on the left*, by the eighth, ninth, and tenth costal cartilages; and *below*, by a line drawn between the lower margins of the tenth costal cartilages. To the left of this area, the anterior wall of the stomach is overlapped by the diaphragm; to the right, it is covered by the liver. Below, the stomach is in relation with the transverse colon. The fundus of the stomach is in relation with the spleen. The esophageal opening of the stomach is placed behind the seventh left costal cartilage, about an inch from the left margin of the sternum. The pyloric orifice of the stomach is situated a little to the right of the middle of a line drawn between the ends of the seventh ribs. The stomach is entirely covered by peritoneum except along the greater and the lesser curvatures and at the posterior aspect of the esophageal opening.

The following **arteries** are distributed to the stomach: the gastric, a branch of the celiac axis, to the lesser curvature; the pyloric, a branch of the hepatic, to the pylorus; the gastro-epiploica dextra, a branch of the gastro-duodenal, to the right side of the greater curvature; the gastro-epiploica sinistra, a branch of the splenic, to the left side of the greater curvature; and the vasa brevia, branches of the splenic, to the fundus.

The **nerves** to the stomach come from the solar plexus and the left pneumogastric nerve. The former branches supply the posterior wall; the latter supply the anterior wall of the organ. These branches form the plexus of Auerbach, in the muscular coat, and the plexus of Meissner, in the submucous coat. (Morris, p. 974; Gray, p. 999.)

THE SMALL INTESTINE.

The small intestine begins at the pyloric orifice of the stomach and extends to the right iliac region to empty into the large intestine, through the ileo-cecal opening. It is disposed in a complicated, coiled manner, principally in the umbilical region. It is surrounded by the large intestine. The small intestine is about twenty-two and one-half feet in length. It is divided into the *duodenum*, the *jejunum*, and the *ileum.*

The **duodenum** is about ten inches long. Beginning at the pyloric extremity of the stomach, the duodenum passes upward to the under surface of the liver; it then bends sharply upon itself and passes downward, in front of the right kidney, to the third lumbar vertebra; it then passes obliquely across the body of the second lumbar vertebra to the left side of the vertebral column, and then passes upward to empty into the jejunum. These portions of the duodenum are known, respectively, as the *ascending,* the *descending,* the *transverse,* and the *second ascending portions.* The duodenum is a retroperitoneal organ. Except for its first, ascending portion, which is partially invested by peritoneum.

RELATIONS.—The ascending portion of the duodenum is in relation, above, with the quadrate lobe of the liver and the neck of the gall bladder; and behind, with the portal vein, the hepatic artery, and the common bile duct. The descending portion of the duodenum is in relation, in front, with the transverse colon; behind, with the right kidney and the inferior vena cava; and to the left, with the pancreas and the common bile duct. The transverse portion of the duodenum is in relation, in front, with the transverse colon and the transverse mesocolon; behind, with the body of the second lumbar vertebra, the inferior vena cava, the thoracic duct, and the abdominal aorta ; above, with the pancreas and the superior mesenteric artery ; and below, with the inferior mesenteric artery. The second ascending portion of the duodenum is in relation with the peritoneum which surrounds it as it passes to join the jejunum.

The **jejunum** is about eight feet in length. The **ileum** is about fourteen feet in length. There is no sharp line of demarcation between these two portions of the small intestine.

The mucous membrane of the small intestine presents numerous transverse folds which are known as the **valvulæ conniventes.** These structures are best marked in the duodenum ; they become less well developed as we pass through the jejunum, and in the ileum they disappear. In the wall of the small intestine we find the glands of Brunner, the crypts of Lieberkühn, the solitary glands, and Pyer's patches.

The **glands of Brunner** are found only in the duodenum. The **crypts of Lieberkühn** and **solitary glands** are found throughout the small intestine. **Pyer's patches** are found only in the ileum, opposite to the attachment of the bowel to the mesentery.

The small intestine receives its **blood supply** from the following vessels : the superior pancreatico-duodenal artery, a branch of the gastro-duodenal, supplies the upper portion of the duodenum. The inferior pancreatico-duodenal artery, a branch of the superior mesenteric, supplies the lower portion of the duodenum. The vasa intestini tenuis, branches of the superior mesenteric, supply the jejunum and the ileum.

The **nerves** are derived from the sympathetic plexus surrounding the superior mesenteric artery. They form the plexus of Auerbach, in the muscular coat of the gut, and the plexus of Meissner, in the submucous coat of the intestine. (Morris, p. 978; Gray, p. 1008.)

THE LARGE INTESTINE.

The **large intestine** is about five feet in length. It is of much larger calibre than the small intestine and surrounds the small bowel on three sides. The opening from the small intestine into the large intestine is known as the **ileo-cecal opening.** It is guarded by two parallel folds of mucous membrane, known as the **ileo-cecal valve.** The serous coat of the large intestine presents numerous small pouches of peritoneum filled with fat, which are known as the **epiploic appendages.**

The large intestine is divisible into the *cecum,* the *colon* and the *rectum.*

The **cecum** is usually entirely surrounded by peritoneum. It is found in the right inguinal region. The **vermiform appendix** arises from its posterior and internal wall. This is the first portion of the large intestine to be developed. It varies in length from one-half inch to ten inches. It may point in one of three directions; downward into the pelvis, upward toward the liver, and obliquely upward toward the spleen. It corresponds in position to the point of crossing of a line drawn between the anterior superior spines of the ilia and the outer border of the right rectus muscle. **McBurney's point** is situated two inches from the anterior superior spine of the ilium, on a line drawn from that spine to the umbilicus. This point also corresponds to the position of the vermiform appendix. The cavity of the vermiform appendix becomes gradually obliterated as age advances.

According to Treves, there are four types of human cecum: in the infantile type, the cecum is funnel-shaped and the appendix arises from the apex of the bowel. In the second type, the cecum presents two equally well developed sacculations and the appendix arises between them. In the third type, the cecum presents a very large right sacculation, while the left sacculation is small. In this type the appendix arises from the posterior and internal wall of the cecum. In the fourth type, the right sacculation is excessive, while the left sacculation atrophies. In this type the appendix seems to arise from the ileo-cecal junction.

The **colon** is divided into the ascending, the transverse and the descending portions. The **ascending colon** begins in the right inguinal region and passes upward through the right lumbar region, into the right hypochondriac region. Here it lies in relation with the inferior surface of the liver and bends upon itself, forming the **hepatic flexure** of the colon. The **transverse colon** begins in the right hypochondriac region, at the hepatic flexure. It passes transversely across the abdomen through the umbilical region to the left hypochondriac region, where it comes in relation with the under surface of the spleen. Here it forms another sharp bend, the **splenic flexure.** The **descending colon** begins in the left hypochondriac region at the splenic flexure. It passes downward through the left lumbar region to the left inguinal region, where it becomes more tortuous, forming the **sigmoid flexure** or the **omega loop.** The ascending colon lies in front of the right kidney. It is usually covered by peritoneum on its anterior two-thirds. The transverse colon lies below the stomach and describes a curve, the convexity of which is directed downward. The transverse portion of the duodenum lies behind it. The descending colon passes in front of the left kidney. The **ogema loop** or **sigmoid flexure** begins in the left inguinal region at the outer margin of psoas magnus muscle. It passes downward into the pelvis along the left wall of that cavity. On reaching the floor of the pelvis, it passes across the midline to the right side. It then ascends along the right wall of the pelvis, crossing the midline as it goes. It then passes from the brim of the true pelvis to the junction of the second and third pieces of the sacrum where the rectum begins.

The **rectum** begins at the junction between the second and third pieces of the sacrum. It passes along the middle of the curve of the sacrum, pierces the levator ani muscle and terminates in the anus. The anus is guarded by two sphincter muscles, the internal and external. The **internal sphincter ani** is composed of involuntary muscle. The **external sphincter** is composed of voluntary muscle.

The large intestine is supplied with **blood** by the following arteries: the ileo-colic, a branch of the superior mesenteric, to the ileo-cecal region; the right colic, a branch of the superior mesenteric, to the ascending colon; the middle colic, a branch of the superior mesenteric, to the transverse colon; the left colic, a branch of the inferior mesenteric, to the descending colon; the sigmoid, a branch of the inferior mesenteric, to the sigmoid flexure; the superior hemorrhoidal, a branch of the inferior mesenteric, to the first portion of the rectum; the middle hemorrhoidal, a branch of the internal iliac, to the middle of the rectum; the inferior hemorrhoidal, a branch of the internal pudic, to the anus.

The **nerves** to the large intestine come from the sympathetic plexuses which surround the arteries of which the nutrient vessels are branches. These nerves form the plexus of Auerbach in the muscular coat and the plexus of Meissner in the submucous coat. (Morris, p. 983; Gray, p. 1027.)

THE LIVER.

The **liver** is a compound tubular gland. It is situated in the right hypochondriac region and in the epigastric region. It has five **lobes :** (1) the *right lobe,* (2) the *left lobe,* (3) the *lobus quadratus,* (4) the *lobus Spigelii,* and (5) the *caudate lobe.* It has five **fissures :** (1) the *fissure for the round ligament,* (2) the *fissure for the ductus venosus,* (3) the *fissure for the gall bladder,* (4) the *fissure for the inferior vena cava,* and (5) the *transverse fissure.*

The **fissure for the round ligament** and the **fissure for the ductus venosus** are collectively spoken of as the **longitudinal fissure.** These fissures divide the right lobe of the liver from the left lobe. The **fissure for the round ligament** is frequently bridged over by a band of liver substance which is known as the **pons hepatis.**

The **quadrate lobe** of the liver is bounded, *on the right,* by the fissure for the gall bladder; *on the left,* by the fissure for the round ligament; *behind,* by the transverse fissure; and *in front,* by the anterior margin of the liver.

The **lobus Spigelii** is bounded, *in front,* by the transverse fissure; *on the right,* by the fissure for the ductus venosus; and *on the left,* by the fissure for the vena cava.

The **caudate lobe** of the liver is a narrow mass of liver substance which extends from the lobus Spigelii to the right lobe of the liver.

The liver is attached to the abdominal parietes by five **ligaments:** (1) the *falciform ligament*, (2) the *coronary ligament*, (3) the *right lateral ligament*, (4) the *left lateral ligament*, and (5) the *round ligament*.

The **falciform ligament** is a double fold of peritoneum which passes from the under surface of the diaphragm to the superior surface of the liver and then passes downward, above the round ligament, to the umbilicus. It passes from before backward on the superior surface of the organ and separates the left lobe from the right lobe on that surface.

The **coronary ligament** is a double fold of peritoneum which extends from side to side, attaching the posterior border of the organ to the diaphragm.

The **lateral ligaments** pass from the right and left limits of the coronary ligament to the side walls of the abdomen. They are composed of peritoneum.

The **round ligament** is the obliterated umbilical vein. It passes from the umbilicus, beneath the peritoneum, to the transverse fissure of the liver. It lies in the fissure for the round ligament.

The liver has three **surfaces:** (1) the *superior surface*, (2) the *inferior surface*, and (3) the *posterior surface*. The **inferior surface** is situated in front of the transverse fissure. The **posterior surface** is placed behind the transverse fissure.

Through the **transverse fissure** of the liver we see the heptic artery, the portal vein, the heptic duct, lymphatics, and nerves passing. The artery lies to the left, the heptic duct lies to the right, and the portal vein lies between and behind these vessels.

RELATIONS.—The superior surface of the liver is in relation with the diaphragm. The heart makes a distinct impression on the left lobe of the organ although they are separated by the diaphragm. The anterior border of the liver crosses the angle between the diverging costal cartilages, obliquely from right to left, from the ninth, right costal cartilage to the eighth left costal cartilage. In the midclavicular line the liver extends from the fifth rib to the costal margin. In the midaxillary line the lower border of the liver reaches the tenth rib. The inferior surface of the liver is in relation with the stomach, the ascending portion of the duodenum, the hepatic flexure of the colon and the right kidney. The stomach is in relation with the left lobe of the liver. The ascending portion of the duodenum lies between the caudate lobe and the neck of the gall bladder. The hepatic flexure of the colon and the right kidney are in relation with the right lobe of the liver; the former is the more anteriorly placed. The posterior surface of the liver is in relation with the ninth and the tenth thoracic vertebræ, the abdominal aorta, the inferior vena cava, the thoracic duct, the esophagus and the right suprarenal body. The fissure for the ductus venosus, the fissure for the inferior vena cava, and the lobus Spigelii are seen on this surface.

The peritoneum completely invests the superior surface of the liver except along the attachment of the falciform ligament. The inferior surface of the liver is covered by peritoneum, except at the attachment of the gall bladder, along the transverse fissure, and at the attachment of the round ligament. The posterior surface of the liver is devoid of peritoneum, except the lobus Spigelii, which is covered by that membrane.

The liver weighs about three pounds (1500 grams); the weight varies between fifty and sixty ounces.

The liver is supplied with nutrient **blood** by the hepatic artery. The portal vein carries blood to the liver which is charged with the products of digestion. It is from this blood that the hepatic cells select the substances which the liver is designed to elaborate. The hepatic veins drain the liver and empty into the inferior vena cava as it lies in its fissure on the posterior surface of the organ.

The **nerves** which supply the liver come from the solar plexus and from the left pneumogastric nerve.

The **gall bladder** is a pear-shaped sac which contains the bile. It is situated in the fissure for the gall bladder on the inferior surface of the right lobe of the liver, to the right of the quadrate lobe. Its fundus points forward and is seen at the anterior margin of the organ, corresponding with the ninth costal cartilage. It empties by the **cystic duct,** through which passage it is also filled. The inferior surface of the gall bladder is covered by peritoneum.

The **hepatic duct** is formed at the transverse fissure of the liver by the union of a duct from the right lobe and a duct from the left lobe of the liver. The hepatic duct, thus formed, then joins with the cystic duct to form the **common bile duct.** The common bile duct passes through the gastro–hepatic omentum, lying to the right of and parallel to the hepatic artery. It then passes behind the first portion of the duodenum and between the head of the pancreas and the descending portion of the duodenum to empty into the descending portion of that division of the small intestine. As the common bile duct passes through the wall of the bowel it presents a dilatation in its lumen which is known as the **ampulla of Vater.** The duct of the pancreas empties into this ampulla. The opening of the common bile duct is marked by a papilla which is placed about four inches from the pylorus of the stomach. The orifice of the common bile duct is much narrower than the ampulla of Vater. (Morris, p. 990; Gray, p. 1047.)

THE PANCREAS.

The pancreas is a racemose gland which is situated in the epigastric and the left hypochondriac regions, close to the posterior abdominal wall. It is composed of a *body*, a *head*, and a *tail*. It weighs about three ounces (90 grams).

RELATIONS.—The head of the pancreas is received into the loop formed by the duodenum. The common bile duct passes between it and the bowel and the portal vein lies behind it. The body of the pancreas is in relation, in front, with the stomach and the ascending layer of the transverse mesocolon: behind, with the body of the first lumbar vertebra, the abdominal aorta, the inferior vena cava, the thoracic duct, and the crura of the diaphragm; below, with the transverse portion of the duodenum from which it is separated by the superior mesenteric artery. The splenic artery and vein pass along the superior border of the gland. The tail of the pancreas crosses the left kidney and is in relation with the spleen.

The pancreas is supplied by the following **arteries:** the small pancreatic and the large pancreatic, branches of the splenic; the superior pancreatico-duodenal, a branch of the gastro-duodenal, and the inferior pancreatico–duodenal, a branch of the superior mesenteric.

The **nerves** which supply the pancreas are branches of the solar plexus.

The **duct of the pancreas** is known as the **canal of Wirsung.** It empties into the descending portion of the duodenum, in common with the common bile duct. (Morris, p. 1001; Gray, p. 1067.)

THE SPLEEN.

The **spleen** is a compound lymph gland. It is situated in the left hypochondriac region. Its axis corresponds with that of the tenth rib. Its anterior border is on a line drawn from the left sterno-clavicular articulation to the tip of the eleventh left rib. Its upper limit is indicated by the eighth rib; its lower limit is indicated by the eleventh rib. The spleen has three **surfaces,** an *anterior surface*, an *internal surface* and a *posterior surface*. The **hilum** of the spleen is the position at which the blood vessels enter and leave the organ. The spleen weighs about six ounces (180 grams).

RELATIONS.—The external surface of the spleen is in relation with the diaphragm and the eighth, ninth, tenth, and eleventh ribs. The anterior surface is in relation with the fundus of the stomach, the tail of the pancreas, and the splenic flexure of the colon. The internal surface is in relation with the left kidney. The spleen is entirely covered by peritoneum, except at the hilum.

The spleen is supplied with **blood** by the splenic artery.

The **nerves** to the spleen come from the solar plexus

and from the right pneumogastric nerve. (Morris, p. 1003; Gray, p. 1073.)

THE DEVELOPMENT OF THE DIGESTIVE SYSTEM.

The **gut tract** is formed by the anterior folding of the splanchnopleure. When the splanchnopleure is united anteriorly we have the gut divided into the **foregut,** the **midgut,** and the **hindgut.**

The **mouth** is formed by a folding in of the surface ectoderm. It is originally separated from the pharynx by a thin partition. This partition soon breaks through and the mouth becomes continuous with the pharynx.

The **pharynx,** the **esophagus,** and the **stomach** are formed from the foregut. The stomach is a dilated portion of the foregut. As it develops, it twists from left to right and from behind forward, so that the left surface of the vertical fetal organ becomes the anterior surface of the adult stomach. This explains why the left pneumogastric nerve lies in front of the stomach.

The **midgut** extends from the fundus of the stomach to the umbilicus. This portion of the primitive gut becomes the **small intestine** in the adult. It grows rapidly and becomes coiled in the central portion of the abdominal cavity. It is early connected to the umbilical vesicle by the umbilical duct. This duct usually atrophies. In some cases it persists, forming **Meckel's diverticulum.**

The **hindgut** extends from the umbilicus to the anus. It develops more slowly than does the midgut, to form the **large intestine.** In its growth it becomes thrown upward and in front of the midgut, explaining the position of the transverse colon in front of the duodenum. The ileo-cecal region, at first, lies beneath the liver; but it rapidly grows downward into the right iliac region. The **vermiform appendix** is the first portion of the large intestine formed. The outer wall of the cecum grows more rapidly than the remainder of that portion of the intestinal tract, throwing the appendix posteriorly and internally. The **anus** is formed by an infolding of the surface ectoderm. It is separated from the remainder of the rectum by a thin partition which soon disappears. In some instances this partition remains, producing an imperforate rectum.

The **peritoneum** is formed by a specialization of the mesothelium which lines the celom.

The **teeth** are developed partly from the ectoderm and partly from the mesoderm. In the mouth of the fetus the ectoderm thickens over the position of the alveolar arches, forming the **dental ridge.** This ridge, which is formed of plugs of epithelium, sinks into the underlying mesoderm forming the **enamel organs** and producing the **dental groove** where, formerly, there was a ridge. As the enamel organ develops, the mesoderm beneath it thickens to form the **dental papilla.** This papilla invaginates the enamel organ so that the latter forms a cap for the former. As growth continues, the enamel organ severs its connection from the overlying ectoderm. Its cells then become differentiated into three layers; the inner layer form the **enamel prisms,** the middle layer disappears, and the outer layer persists as a delicate, cuticular covering, the **membrane of Nasmyth.** The dental papilla becomes furnished with blood vessels and nerves, and from it the **dentine,** the **pulp,** and the **cementum** are formed. The connective tissue cells which form the dentine are known as **odontoblasts.** The oldest enamel and the oldest dentine lie in close apposition. In the course of development, the enamel organ sends off a process into the surrounding mesoderm which forms the **enamel organ for the permanent tooth.** This process is at first connected to the primary enamel organ by a narrow band of epithelial tissue. This tissue is finally absorbed and the permanent tooth develops in the same manner as does the temporary tooth.

The **liver** grows from the entoderm lining the midgut, as solid plugs of cells. These plugs divide and subdivide and form the lobules and the ducts of the organ.

The **pancreas** grows from the entoderm lining the midgut.

The **salivary glands** grow from the ectoderm of the oral cavity. The plugs of epithelium are at first solid; but they rapidly present a lumen and form the acini and the ducts of the glands. (Quain, p. 99; A. T. O., p. 112; Piersol, p. 149.)

CHAPTER XVI.

THE GENITO-URINARY SYSTEM.

THE KIDNEYS.

The **kidneys** are compound tubular glands, situated on either side of the vertebral column. Each kidney weighs about four and one-half ounces (150 grams). They are retroperitoneal organs and rest upon the diaphragm and the anterior layer of the lumbar fascia. They are surrounded by a variable amount of **perirenal fat**. At the inner border of the kidney there is an opening which allows the passage of the blood vessels and duct to and from the substance of the organ; this opening is known as the **hilum** of the kidney. Just inside the hilum of the kidney, there is a dilated fossa which contains the blood vessels and the pelvis of the kidney, known as the **sinus** of the kidney. The structures which pass through the hilum of the kidney are: the renal vein, the renal artery, and the ureter. These structures are here enumerated in their usual order from before backward.

The **right kidney** is situated principally in the right lumbar region; but it projects somewhat into the right hypochondriac, the epigastric, and the umbilical regions. It extends from the lower border of the eleventh rib to the transverse process of the third lumbar vertebra.

RELATIONS.—In front, it is in relation with the inferior surface of the liver, the hepatic flexure of the colon, and the descending portion of the duodenum. Behind, it is in relation with the diaphragm, the anterior layer of the lumbar fascia, the psoas magnus muscle, the last thoracic, and the ilio-inguinal and ilio-hypogastric nerves. Above, it is in relation with the right suprarenal body and is separated from the pleura by the diaphragm.

The **left kidney** extends from the upper border of the eleventh rib to the transverse process of the second lumbar vertebra.

RELATIONS.—In front, it is in relation with the stomach, the tail of the pancreas, the splenic flexure of the colon, a small portion of the spleen, and the splenic artery and vein. Behind, it is in relation with the diaphragm, the anterior layer of the lumbar fascia, the psoas magnus muscle, the last thoracic, and the ilio-inguinal and ilio-hypogastric nerves. Above, it is in relation with the left suprarenal body and is separated from the left pleura by the diaphragm. (Morris, p. 1020; Gray, p. 1127.)

THE URETERS.

The **ureter** is the excretory duct of the kidney. It is about twelve inches long, extending from the hilum of the kidney to the bladder.

The ureter begins by a dilated extremity or **pelvis** which is contained in the sinus of the kidney. Two pelves are usually described, the **superior pelvis** and the **inferior pelvis,** into each of which several small tubes, **infundibula,** empty. At the upper extremity of each infundibulum there is a dilated pouch, called the **calyx,** into which a renal papilla projects. The **renal papillæ** are the terminations of the Malpighian pyramids.

RELATIONS.—From the hilum of the kidney, the ureter passes downward, behind the peritoneum and above the psoas magnus muscle, to the point of bifurcation of the common iliac artery, which it crosses. It then passes down into the true pelvis and lies on the side wall of that cavity until it reaches the posterior surface of the bladder. It passes obliquely through the wall of the bladder to empty into that organ at either extremity of the base of the vesical trigone. In its passage through the abdomen, the ureter is crossed on either side by the spermatic artery and vein in the male, and by the ovarian artery and vein in the female. The **right ureter** lies to the right of the inferior vena cava and beneath the ileo-cecal region.

The **left ureter** lies to the left of the abdominal aorta and beneath the omega loop of the colon. In the pelvis, in the male, the ureter is crossed by the vas deferens and, on the posterior wall of the bladder, lies to the outer side of that structure and to the inner side of the seminal vesicle. In the female, the ureter passes through the lower portion of the broad ligament and then lies by the side of the cervix of the uterus and of the upper portion of the vagina. (Morris, p. 1029; Gray, p. 1136.)

The kidney is supplied with arterial **blood** from the renal artery, which is a branch of the abdominal aorta. Since the abdominal aorta lies to the left of the median line of the abdomen, the right renal artery is somewhat longer than the corresponding vessel of the left side. The renal artery enters the kidney by passing through the hilum, and then divides into numerous branches. These branches are given off in the sinus of the kidney. They then enter the substance of the kidney and pass through the columns of Bertini to the line of junction between the cortex and the medulla. Here they form short arches which do not anastomose, and which lie parallel with the long axis of the kidney. From these arches branches, known as the **arteriæ rectæ,** pass down into the Malpighian pyramids, forming capillary networks around the uriniferous tubules. Branches pass from the arterial arches into the cortex, as the interlobular cortical arteries. As these vessels pass outward they give off branches which form the **glomeruli** of the Malpighian bodies. Each glomerulus has an **afferent vessel** and an **efferent** vessel, between which is a highly convoluted plexus of capillaries. The efferent vessel contains arterial blood; it passes to the uriniferous tubules of the cortex around which it forms a network.

The **veins** begin beneath the capsule of the kidney in stellate shaped groups, **stellate veins,** and pass with other veins from the cortex, **interlobular veins.** The interlobular veins are joined by the **venæ rectæ** of the medulla, and form large vessels which, in the sinus of the kidney, unite to form the renal veins. The renal vein leaves the kidney by passing through the hilum and empties into the inferior vena cava. The left renal vein is longer than the right because the inferior vena cava is to the right of the median line. (Morris, p. 1026; Gray, p. 1133.)

THE URINARY BLADDER.

The **bladder** is a musculo-membranous pouch, composed of a fibrous coat, a muscular coat, and a mucous coat. The fibrous coat of the bladder is derived from the vesical layer of the recto-vesical fascia and, on the posterior wall of the organ, there is, in addition, a serous coat derived from the peritoneum.

The bladder is situated behind the symphisis pubis and, when empty, is contained within the cavity of the true pelvis. When full, however, it projects above the superior opening of the pelvis and occupies the hypogastric region of the abdomen.

The bladder is attached to the abdominal walls and to the surrounding viscera by two sets of **ligaments,** the *true* and the *false*. The **true ligaments** are five in number and, with one exception, are derived from the recto-vesical fascia. They are : (1 and 2) the *anterior*, (3 and 4) the *lateral*, and (5) the *superior.*

The **anterior true ligaments** are formed by folds of the recto-vesical fascia which pass from the anterior wall of the bladder to the prostate gland and thence to the pubic bones, to which they are attached. They are called the **pubo-prostatic ligaments.**

The **lateral true ligaments** are folds of the recto-vesical fascia which pass from the lateral aspect of the bladder to the fascia covering the levator ani muscle.

The **superior true ligament** passes from the superior wall of the bladder to the umbilicus. It is the obliterated allantoic duct (see page 12) and is called the **urachus.**

The **false ligaments** are also five in number and are formed by folds of peritoneum passing from the bladder to adjacent structures. They are : (1 and 2) the *lateral*, (3 and 4) the *posterior*, and (5) the *superior.*

The **lateral false ligaments** pass from the lateral aspect of the bladder to the side wall of the pelvis, above the lateral true ligaments.

The **posterior false ligaments** pass, in the male, from the wall of the rectum to the posterior wall of the bladder. They are known as the **recto-vesical ligaments**. In the female, these ligaments are very short and, instead of coming from the rectum, come from the sides of the uterus.

The **superior false ligament** is the fold of peritoneum which surrounds the urachus.

RELATIONS.—The anterior surface of the bladder is in relation with the symphisis pubis, being separated from it by a small space which is occupied by areolar tissue; this is known as the **space of Retzius.** When the bladder is distended this space is pushed upward and lies between the bladder and the lower portion of the abdominal wall, giving an area through which operations on the viscus may be performed without injuring the peritoneum. The posterior surface of the bladder is covered by peritoneum and, in the male, is separated from the anterior wall of the rectum by coils of small intestine, which occupy the recto-vesical pouch of the peritoneum. In the female, the posterior surface of the bladder is in close relation with the anterior wall of the uterus, which rests upon it. The sides of the bladder are crossed, in the male, by the vas deferens, on its way to the fundus of the organ. It is only partially covered by peritoneum. The ureter crosses this surface of the bladder in both sexes.

The **fundus** of the bladder is the most dependent part of the organ. In the male it is in close relation with the anterior wall of the rectum and presents, on either side of the median line, the vas deferens, internally, the seminal vesicle, externally, and the ureter between the two, piercing the wall of the organ.

In the female, the fundus of the bladder is in relation with the cervix of the uterus and the upper portion of the anterior wall of the vagina.

The mucous coat of the bladder is thrown into folds or **rugæ** in all parts of the organ except over a small area at the fundus. This smooth area is known as the **vesical trigone.** It is bounded, *above*, by the orifices of the two ureters, and *below*, by the orifice of the urethra. There is a small projection of the mucosa of the bladder into the orifice of the urethra which is called the **vesical uvula.** The bladder is lined by transitional epithelium.

The bladder is supplied with **blood** by the superior, middle, and inferior vesical arteries, branches of the internal iliac artery. The superior vesical artery is the pervious portion of the obliterated hypogastric artery. (Morris, p. 1031; Gray, p. 1139.)

THE URETHRA.

The **urethra** is the name given to the canal by which the urine is passed away from the bladder. In the male it is about seven and one-half inches in length; in the female about one inch and one-quarter in length.

The **male urethra** is divisible into three parts; the *prostatic*, the *membranous*, and the *spongy* or *penile*.

The **prostatic portion** of the male urethra passes through the prostate gland. It is about one inch in length. On the floor of this portion of the urethra there is a central crest of mucous membrane, known as the **urethral crest** or **verumontanum.** In the centre of this crest there is the opening of a blind pouch, the **sinus pocularis** or **uterus masculinus,** which extends backward into the fissure of the prostate gland. In the urethral crest on either side of the opening of the uterus masculinus the **opening of the ejaculatory duct** may be seen. On either side of the urethral crest there is a depression known as the **prostatic sinus** in the bottom of which the openings of the ducts of the prostate gland are to be seen. The prostatic portion of the urethra is lined by transitional epithelium.

The **membranous portion** of the urethra is about one-half inch in length; it lies between the superior and the inferior layers of the triangular ligament of the perineum. It is surrounded by the compressor urethræ muscle and is in relation, on either side, with Cowper's glands. It is lined by stratified columnar epithelium.

The **spongy portion** of the urethra is about six inches in length. It passes through the spongy body of the penis to terminate, as the **external urinary meatus,** on the glans penis. In front of the inferior layer of the triangular ligament, the calibre of the urethra is larger than it is in the remainder of its extent. From this fact and from the fact that the spongy body of the penis here presents a bulbous enlargement, this portion of the spongy urethra is known as the **bulbous urethra.** The bulbous urethra is limited anteriorly by Colles' fascia. The junction of the bulbous urethra and the membranous urethra is termed the **bulbo-membranous junction.** The ducts of Cowper's glands empty into the bulbous urethra. The remainder of the spongy urethra presents the orifices of the **glands of Littré.** One of these orifices, larger than the others, situated on the superior wall of the urethra, about an inch behind the external urinary meatus, is called the **lacuna magna.** Just behind the external urinary meatus the spongy urethra presents a dilatation which is known as the **fossa navicularis.** The fossa navicularis is lined by stratified squamous epithelium; the remainder of the spongy urethra is lined by simple columnar epithelium.

The **female urethra** passes between and pierces the anterior and posterior layers of the triangular ligament of the perineum. It opens into the vestibule. It is in relation with the vagina by its posterior wall. It is lined by stratified squamous epithelium. (Morris, pp. 1050 and 1056; Gray, pp. 1146 and 1167.)

THE MALE GENERATIVE ORGANS.

The **scrotum** is a membranous pouch which contains the testicles. It is composed of the following tissues: (1) the *skin,* (2) the *dartos,* (3) the *external spermatic fascia,* (4) the *middle spermatic fascia,* (5) the *internal spermatic fascia,* and (6) the *tunica vaginalis.*

The **skin** of the scrotum presents a central median raphé and is usually found in numerous marked folds.

The **dartos** is the superficial fascia of the scrotum. It is continuous with the superficial fascia of the abdomen and of the thighs and contains involuntary muscle fibres.

The **external spermatic fascia** is continuous with the intercolumnar fascia (see p. 149).

The **middle spermatic fascia** is continuous with the cremaster muscle (see p. 150).

The **internal spermatic fascia** is continuous with the infundibuliform fascia (see p. 151).

The **tunica vaginalis** is a process of peritoneum which is carried downward into the scrotum with the descent of the testicle. Only the parietal layer of the tunica vaginalis is to be considered as one of the layers of the scrotum. The **visceral layer of the tunica vaginalis** closely invests the testicle. It is prolonged into the groove between the testicle and the epididymis, thus forming the **digital fossa.** If the testicle is held by the spermatic cord the digital fossa will point toward the side to which the organ belongs.

The **spermatic cord** is the structure which carries the vessels and nerves to, and the excretory duct from the testicle. It is composed of (1) the *spermatic artery,* (2) the *spermatic plexus of veins,* (3) the *spermatic plexus of the sympathetic system,* (4) the *vas deferens,* (5) the *artery of the vas deferens,* (6) the *cremaster muscle,* (7) the *artery to the cremaster muscle,* (8) the *genital branch of the genito-crural nerve,* and (9) the *obliterated portion of the tunica vaginalis.* It begins at the internal abdominal ring and ends at the testicle, passing through the inguinal canal.

The scrotum is supplied with **blood** by the superficial external pudic and the deep external pudic arteries, branches of the common femoral; the superficial perineal arteries, branches of the internal pudic; and twigs from the artery to the cremaster muscle.

The following **nerves** supply the scrotum: the genital branch of the genito-crural nerve, the inguinal branch of the ilio-inguinal nerve, the superficial perineal branches of the internal pudic nerve, and the inferior pudendal branch of the small sciatic nerve. (Morris, pp. 1038 and 1045; Gray, p. 1153.)

THE TESTICLE.

The sexual gland of the male is known as the **testicle.** The testicle is a compound tubular gland which is contained in the scrotum. It is invested by a serous pouch known as the tunica vaginalis. It weighs about one ounce (24.5 grams). The left testicle is usually placed somewhat lower in the scrotum than is its fellow of the right side. The spermatic cord is attached to its posterior border.

The **gubernaculum testis** is a band of involuntary muscle which is attached to the scrotum by one extremity and to the testicle by its other extremity. This tissue is also attached to the pillars of the external abdominal ring.

The **epididymis** is a convoluted tubule, about one inch in length, which is attached to the posterior border of the gland. It is composed of a **globus major** or **head,** above; a **body ;** and a **globus minor** or **tail,** below. The head of the epididymis receives the vasa efferentia from the testicle. Between the head of the epididymis and the upper border of the testicle there is a depression which is known as the **digital fossa** of the testicle. The tunica vaginalis is prolonged into this groove.

The **hydatid of Morgagni** is a small, sessile body which is attached to the head of the epididymis. It is the remains of the upper extremity of the Müllerian duct. The **stalked hydatid** is attached to the globus major of the epididymis. It is developed from the anterior portion of the Wolffian body.

The excretory duct of the testicle is known as the **vas deferens.** The vas deferns begins at the tail of the epididymis and passes upward as one of the constituents of the spermatic cord, through the inguinal canal, and out of the internal abdominal ring. It then passes across the deep epigastric artery and runs downward, across the external iliac artery and vein, and along the side wall of the pelvis to reach the posterior surface of the bladder. On the posterior surface of the bladder it crosses the ureter and lies to the inner side of the seminal vesicle. It joins with the duct of the seminal vesicle to form the **ejaculatory duct.** Just before it joins with the duct of the seminal vesicle, the vas deferens presents a dilatation which is known as the **ampulla.** The ureter pierces the posterior wall of the bladder between the vas deferens and the seminal vesicle. The vas deferens has three coats; a mucous, a muscular, and a fibrous. It can be told from the other structures which compose the spermatic cord by the thickness of its walls, which gives it a wiry feel.

The **vas aberrans** is a narrow tubule, ending in a blind extremity, which is found between the epididymis and the vas deferens. At times it opens into the epididymis and frequently it is found springing from the vas deferens.

The **seminal vesicle** is a convoluted tubule which is found on the posterior wall of the bladder, external to the vas deferens. It serves for the reception of the semen. It is connected to the vas deferens by the duct of the seminal vesicle.

The **ejaculatory duct** is formed by the union of the duct of the seminal vesicle and the vas deferens. It passes through the fissure of the prostate gland and empties into the floor of the prostatic urethra. The orifice of the ejaculatory duct may be seen on the urethral crest by the side of the uterus masculinus. (Morris, p. 1038; Gray, p. 1156.)

The **prostate gland** is a racemose gland which is situated just in front of the bladder and behind the superior layer of the triangular ligament of the perineum. It weighs about four and one-half drams (18 grams). It consists of a median and two lateral lobes. The two lateral lobes meet in front of the urethra and are separated from the median lobe, which lies behind the urethra by the prostatic fissure. The ejaculatory ducts and the uterus masculinus are found in this fissure.

RELATIONS.—By its anterior surface, the prostate gland is in relation with the symphisis pubis. Posteriorly, it is in relation with the rectum, through the wall of which it can be felt. Laterally, it is in relation with the levator ani muscles. The base of the gland is in relation with the bladder and the apex rests against the superior layer of the triangular ligament of the perineum. The prostatic portion of the urethra passes through it.

There is a large amount of muscular tissue between the acini of the gland. (Morris, p. 1036; Gray, p. 1148.)

The **sinus pocularis** or **uterus masculinus** is a blind pouch, which is the remains of the lower extremity of the Müllerian duct. It opens into the prostatic portion of the urethra and its blind extremity is found in the prostatic fissure. (Morris, p. 1050; Gray, p. 1146.)

The **penis** is an organ which is composed of three rounded, muscular bodies. The two **corpora cavernosa** lie side by side on the dorsal surface of the organ. They are composed of erectile tissue and end posteriorly as diverging muscular masses, the **crura,** which are attached to the rami of the pubes and the ischium. Anteriorly, the corpora cavernosa are conical and serve for the support of the glans penis.

The **corpus spongiosum** begins, posteriorly, as an expanded **bulb** which is found between the inferior layer of the triangular ligament of the perineum and Colles' fascia. Anteriorly, it expands into a heart-shaped extremity, the **glans penis,** which is supported by the anterior, conical ends of the corpora cavernosa. The position of union of the glans penis and the anterior extremities of the corpora cavernosa is known as the **neck of the penis**. Just in front of the neck, the glans penis presents a flaring edge which is known as the **corona glandis.**

The urethra passes through the corpus spongiosum and ends at the anterior extremity of the glans penis as the external urinary meatus.

The two cavernous bodies are separated from each other by a fibrous partition which is known as the **pectiniform septum.** The whole organ is enveloped by a layer of fibrous tissue, which is known as **Buck's fascia,** and this is, in turn, covered over by skin. Anteriorly, the skin is reflected from the neck of the penis, forward, over the glans penis as the **prepuce.** The prepuce is attached to the ventral surface of the glans by a slip of the integument, known as the **frenum of the prepuce.** The inner aspect of the prepuce is well supplied with modified sebaceous glands which are spoken of as the **odoriferous glands of Tyson.**

The dorsal aspect of the penis is attached to the under surface of the symphisis pubis by a dense, fibrous band, the **suspensory ligament of the penis.** In front of the suspensory ligament, the penis bends downward, in the flaccid condition, to form the **penile angle.**

The penis is supplied with **blood** by the dorsal artery of the penis, the artery of the bulb, and the artery of the corpus cavernosum, branches of the internal pudic artery. The superficial external pudic artery, from the common femoral, also sends twigs to the organ.

The **nerves** to the penis are the superficial perineal nerves and the dorsal nerve of the penis, branches of the internal pudic. Branches of the hypogastric plexus of the sympathetic system supply the corpora cavernosa. (Morris, p. 1046; Gray, p. 1150.)

Cowper's glands are two small, racemose glands which are situated between the superior and the inferior layers of the triangular ligament of the perineum. They are found, one on either side of the membranous urethra. Their ducts pierce the inferior layer of the triangular ligament and empty into the bulbous urethra. (Morris, p. 1082; Gray, p. 1150.)

THE FEMALE ORGANS OF GENERATION.

The **female organs of generation** may be divided into an *external group* and an *internal group.*

The **external genitals** are spoken of as the **vulva** or **pudendum.** This term comprehends the *mons Veneris,* the *labia majora,* the *labia minora,* the *clitoris,* the *vestibule* and the *hymen.*

The **mons Veneris** is a pad of fat which covers the symphisis pubis. The skin which covers it is thickly supplied with crisp hairs.

The **labia majora** are two folds of skin and fascia which guard the vulvar orifice. They extend from the mons Veneris to a point about one inch from the anus. They correspond to the scrotum in the male.

The **labia minora** are two folds of modified skin which are seen inside the labia majora. They begin, just above the clitoris, by a common fold which forms the prepuce of that organ and extend backward to be connected by the **fourchette,** just in front of the posterior extremity of the labia majora.

The **clitoris** is an organ, composed of erectile tissue, which corresponds with the penis in the male. It is situated beneath the anterior extremities of the labia minora, which meet above it in a cutaneous fold which is termed the **prepuce of the clitoris.** It is composed of two **cavernous bodies,** which are attached to the rami of the pubes and the ischium by the **crura** of the clitoris. The free extremity of the clitoris presents the **glans clitoridis.** The erector clitoridis muscle is connected to the organ on either side and it is suspended from the pubic arch by a **suspensory ligament.** It is not traversed by the urethra.

The **vestibule** is the triangular space which is bounded, *in front*, by the clitoris; *laterally*, by the labia minora; and *posteriorly*, by the orifice of the vagina. Beneath the mucous membrane of the vestibule on either side, just within the position of the labia minora, are two masses of erectile tissue which correspond to the divided corpus cavernosum of the penis. They are known as the **bulbi vestibuli.** The vestibule presents the orifice of the urethra, and the ducts of the glands of Bartholin empty into the space.

The **glands of Bartholin** are two racemose glands which are situated behind the bulbi vestibuli, on either side of the vestibule. The ducts of these glands empty into the vestibule. They correspond to the glands of Cowper in the male; but are situated in front of the inferior layer of the triangular ligament.

The vaginal orifice is guarded by a thin fold of mucous membrane which is termed the **hymen.** This membrane does not form a complete septum, but is variously perforated to permit of the exit of the vaginal secretions and the menstrual discharges. In women who have borne children the remains of the hymen surround the vaginal orifice as small, knob-like masses which are called the **carunculæ myrtiformes**. In the female the **fossa navicularis** is the space between the fourchette and the posterior border of the vaginal orifice. (Morris, p. 1053; Gray, p. 1163.)

The **internal genitalia** of the female are the *ovaries*, the *uterus*, the *Fallopian tubes*, and the *vagina*.

The **ovaries** are the female sexual glands. They are situated between the folds of the broad ligament of the uterus and project into the true pelvis from its posterior surface. If the broad ligament is examined from its anterior aspect the ovary is entirely concealed from view. Each ovary weighs about one hundred grains (7 grams).

The ovary is placed in the true pelvis in such a position that its surfaces look inward and outward, and its long axis is in a nearly vertical plane. The borders of the organ would then be designated as anterior and posterior. The Fallopian tube passes along the anterior border and over the superior extremity of the ovary, and lies, by its fimbriated extremity, in relation with the internal or mesial surface and partly with the posterior border. The **fimbria ovarica** is attached to the superior extremity of the anterior border of the ovary. The external or lateral surface of the ovary is in relation with the side wall of the pelvis and is contained in the ovarian groove. The **ovarian groove** is bounded, *above*, by the obliterated hypogastric artery; and *below*, by the ureter.[1]

The ovary is supplied with **blood** by the ovarian artery. The ovarian veins form a plexus in front of the hilum, which is called the **pampiniform plexus.** The ovarian plexus of the sympathetic system supplies the organ with nervous impulses. (Morris, p. 1066; Gray, p. 1175.)

The **uterus** is a hollow muscular organ which is situated in the pelvis, between the rectum, behind, and the bladder, in front. It is composed of a *body*, a *fundus*, and a *cervix*. The **fundus** of the uterus is at the superior extremity of the organ. The **body** of the uterus has an anterior surface, which is partly covered by peritoneum, and a posterior surface, which is decidedly more convex than the anterior surface and which is completely covered by peritoneum. From the superior portion of the lateral aspect of the body of the uterus the Fallopian tubes can be seen passing toward the ovaries. The **cervix** of the uterus (**cervix uteri**) is the narrow, elongated portion of the uterus which projects into the vagina. The anterior wall of the cervix lies entirely without the limits of the peritonium. The posterior wall, on the contrary, is partly covered by peritoneum. The wall of vagina is attached to the posterior wall of the cervix at a higher point than to the anterior wall of the cervix. The cavity of the uterus may be divided into the *cavity of the body* and the *cavity* or *canal of the cervix* of the organ. The **cavity of the body of the uterus** is triangular in shape. The base of the triangle is directed upward and the angle at either end of the base is termed the **cornu ;** it contains the orifice of the corresponding Fallopian tube. The apex of the cavity of the body of the uterus is directed downward and empties into the cervical canal through a constricted **internal os.** The **cervical canal** is about one inch long. It terminates in the vagina by the **external os** or the **os uteri.**

The uterus is normally in the position of slight anteflexion, that is, tilted somewhat forward, with the anterior surface of the body of the organ resting upon the bladder, when the latter organ is empty. Usually there are no coils of intestine interposed between the uterus and the bladder. The uterus, however, is capable of movement and may occupy a different position if the bladder is full and the rectum empty.

The uterus is supported in its normal position by certain liga_ ments. The **ligaments** of the uterus are of two kinds: the *peritoneal ligaments* and the *muscular ligaments.* These ligaments are all paired.

The **peritoneal ligaments** are: (1) the *lateral* or *broad,* (2) the *utero-vesical,* and (3) the *utero-rectal.*

The **muscular ligaments** are: (1) the *utero-inguinal,* or *round,* (2) the *utero-ovarian,* (3) the *utero-sacral,* and (4) the *utero-pelvic.*

The **lateral** or **broad ligaments** of the uterus, one on either side, are double folds of peritoneum, with connective tissue between them, which pass from the sides of the uterus to the side wall of the pelvis. The internal border is attached to the body of the uterus; the superior border lies free in the pelvis; the external border is attached to the obturator fascia; and the inferior border is attached to the recto-vesical fascia. Beyond the fimbriated extremity of the Fallopian tube, the superior border of the broad ligament forms a sharp angle and then passes to the side wall of the pelvis, forming the **infundibulo-pelvic ligament.** This ligament transmits the ovarian artery and vein.

Between the layers of the broad ligament we find: (1) the *ovary,* (2) the *Fallopian tube,* (3) the *utero-ovarian ligament,* (4) the *utero-inguinal ligament,* (5) the *utero-pelvic ligament,* (6) the *parovarium,* (7) the *paroöphoron,* (8) the *stalked hydatid,* (9) *involuntary muscle,* and (10) the *ovarian, uterine,* and *vaginal vessels* and *nerves.*

The **utero-vesical ligaments** are double folds of peritoneum which pass from the uterus to the bladder.

The **utero-rectal ligaments** are double folds of peritoneum which pass from the uterus to the rectum.

The **utero-inguinal,** or **round ligament,** is a rounded bundle of involuntary muscle which, starting from the superior portion of the lateral aspect of the uterus, just below the Fallopian tube, passes through the broad ligament, through the inguinal canal, and the external abdominal ring, to fade away in the tissue of the labium majus. In its course, it passes in front of the internal iliac vessels and across the deep epigastric artery. In the inguinal canal it is surrounded by a reflection of the peritoneum which is known as the **canal of Nuck.**

The **utero-ovarian ligament** extends from the uterus to the anterior border of the ovary.

The **utero-sacral ligament** passes backward, between the folds of the utero-rectal ligament, to be attached to the sacrum.

The **utero-pelvic ligament** is a radiating mass of involuntary muscle which lies between the folds of the broad ligament. It attaches the uterus to the obturator fascia.

The **parovarium** or **epoöphoron** is composed of a few vertically placed tubules which are connected to a single, horizontally placed tubule. It represents parts of the Wolffian body and the Wolffian duct of the fetus.

The **paroöphoron** is the remains of a portion of the Wolffian body.

The **stalked hydatid** is attached to the fimbriated extremity of the Fallopian tube. It represents the remains of fetal structures. (Morris, p. 1058; Gray, p. 1168.)

The **Fallopian tube** begins at the superior portion of the lateral aspect of the body of the uterus and passes outward in the superior border of the broad ligament to end, in relation with the ovary, by a fringed end which is known as the **fimbriated extremity** of the Fallopian tube. All the fimbria lie free in the pelvic cavity except one, the **fimbria ovarica,** which is attached to the superior extremity of the anterior border of the ovary. The Fallopian tube passes along the anterior border of ovary, arches over its superior extremity, and ends in relation with the mesial surface and the posterior border of the organ. In the midst of the fimbria, at the fimbriated end of the Fallopian tube, the **ostium abdominale** is found. This is the entrance into the lumen of the tube. The inner extremity of the Fallopian tube empties into one of the cornua of the uterus. (Morris, p. 1065; Gray, p. 1174.)

The **vagina** is a musculo-membranous tube which passes from the uterus to the vulva, piercing the urethral triangle in its passage. Its posterior wall is in relation with the rectum behind and, in its upper portion receives a reflection of peritoneum. Its anterior wall is in relation with the base of the bladder and with the urethra. The lateral walls are in relation with the levatores ani muscles and, in the upper fourth, with the ureters. The mucous membrane of the vagina is attached to the posteror wall of the cervix uteri at a higher level than it is attached to the anterior wall of the cervix. Between the posterior wall of the cervix and the posterior wall of the vagina there is a pouch which is known as the **vaginal cul=de=sac** (see p. 7). The vagina is about three and one-half inches long on its posterior wall and two and one-half inches long on its anterior wall. It passes upward and backward forming an angle of about 10° with the long axis of the body. The cervix uteri projects into the superior extremity of the vagina. The inferior extremity of the vagina is closed by the hymen. (Morris, p. 1056; Gray, p. 1167.)

THE SUPRARENAL BODIES.

The **suprarenal bodies** are attached to the upper and inner aspect of the kidneys. The **right suprarenal body** resembles a liberty-cap in shape. It is in relation with the posterior surface of the liver.

The **left suprarenal body** is semilunar in outline. It is in relation with the posterior wall of the stomach.

The following **arteries** supply the suprarenal bodies: the suprarenal artery, a branch of the abdominal aorta, and branches from the renal and the phrenic arteries.

The **nerves** to the suprarenal body are very numerous; they form the suprarenal plexus. The twigs which form this plexus come from the solar, the phrenic, and the renal plexuses (see p. 83). (Morris, p. 1028; Gray, p. 1137.)

THE DEVELOPMENT OF THE GENITO-URINARY SYSTEM.

The excretory organ of the embryo is known as the **Wolffian body.** This body is developed in the mesoderm which lines the pleuro-peritoneal cavity and is composed of tubules. The secretion of this body is carried off by the **Wolffian duct** and emptied into the cloaca. The Wolffian body may be divided into an *anterior segment,* a *middle segment,* and a *posterior segment.* In the human subject the anterior and posterior segments remain rudimentary, while the middle segment or **mesonephros** becomes of great embryologic importance.

The **kidney (metanephros)** develops as an outgrowth from the Wolffian duct. This outgrowth forms the **ureter,** then expands to form the **pelvis of the kidney,** and subdivides to form the **calices.** The uriniferous tubules grow as further elaborations from the primary tubule. The connective tissue and blood vessels grow in from the surrounding mesoderm, the glomeruli invaginating the upper extremities of the tubules to form the **capsules of Bowman.**

The **bladder** is developed as a dilatation of the allantoic stalk. The urethra is developed as a channel from the allantoic stalk to the surface. The **urachus** is the obliterated portion of the allantoic stalk from the fundus of the bladder to the umbilicus.

The sexual gland grows independently of the tubules which are developed to carry off its secretion. The **ovary,** in the female, and the **testicle,** in the male, develop from the mesothelium which is internal to the Wolffian body. The two organs have a similar appearance up to the third month; at which time the differentiation into ovary or testicle takes place.

After the kidney is formed a second tube, the **Müllerian duct,** grows beside the Wolffian duct. So that in both sexes the same preliminary steps are taken.

In the female the Müllerian duct persists.

The **Fallopian tubes** and their fimbriated extremities are developed from the upper portions of the Müllerian duct.

The lower portions of the Müllerian ducts fuse to form the **uterus** and the **vagina.** In the female the Wolffian duct atrophies. If it is patulous it is known as **Gartner's duct.** The **parovarium** or **epoöphoron** is developed, partly from the tubules of the Wolffian body (vertical tubules), and partly from the Wolffian duct (horizontal tubule). The **paroöphoron** is the remains of some of the tubules of the posterior segment of the Wolffian body. The **stalked hydatid** is the remains of the anterior segment of the Wolffian body.

In the male, the Wolffian duct and the tubules of the mesonephros persist and form the **vasa efferentia, the coni vasculosi,** the **epididymis,** the **seminal vesicles** and the **vas deferens.**

The **paradidymis** is the remains of the posterior segment of the Wolffian body. The **stalked hydatid** is the remains of the anterior segment of the Wolffian body. The Müllerian duct usu-

ally atrophies except at its superior extremity which forms the
sessile hydatid; and at its inferior extremity, which persists as
the **uterus masculinus.** When the Müllerian duct is patulous
in the male it is known as **Rathke's duct.**

The **external genitals** in both sexes develop from structures which present the same appearance until about the third
month, when differentiation into the characteristics of the sex
takes place. The **genital tubercle** is surrounded by the
genital folds from which it is separated by the **genital
groove.** Outside the genital folds we find the **genital
ridges.**

In the male the genital tubercle gives rise to the **cavernous bodies of the penis** and the **glans penis**; the genital
groove forms the **penile urethra**; the genital folds form the
spongy body of the penis; and the genital ridges become
the **scrotum.**

In the female the genital tubercle gives rise to the **clitoris**
and the **glans clitoridis**; the genital groove forms the **vulvar
orifice;** the genital folds become the **labia minora;** and the
genital ridges form the **labia majora.** (Quain, p. 115; A.
T. O., p. 118.)

TABLE OF ORIGIN OF THE SUBDIVISIONS OF THE GENITO-URINARY TRACT.

Mesoderm.	Kidney, Ureter, Testicle, Ovary, Epididymis, Vas Deferens, Seminal Vesicles, Fallopian Tubes, Uterus, Vagina.
Entoderm.	Bladder and Urethra.

TABLE OF DEVELOPMENT OF INTERNAL GENITAL ORGANS.

WOLFFIAN DUCT.		MÜLLERIAN DUCT.	
Male.	*Female.*	*Male.*	*Female.*
Vasa Efferentia, Coni Vasculosi, Epididymis, Vas Deferens, Seminal Vesicle.	Parovarium or Epoöphoron. If patulous is known as Rathke's Duct.	Sessile Hydatid, Uterus Masculinus. If patulous is known as Gartner's Duct.	Fallopian Tubes, Uterus, Vagina.

TABLE OF DEVELOPMENT OF EXTERNAL GENITALS.

Structure.	*Male.*	*Female.*
Genital tubercle.	Corpora cavernosa and glans penis.	Clitoris and glans clitoridis.
Genital groove.	Penile urethra.	Vulvar orifice.
Genital folds.	Spongy body of penis.	Labia minora.
Genital ridges.	Scrotum.	Labia majora.